Lecture Notes in Mathematics 1683

Editors:
A. Dold, Heidelberg
F. Takens, Groningen

Springer
Berlin
Heidelberg
New York
Barcelona
Budapest
Hong Kong
London
Milan
Paris
Santa Clara
Singapore
Tokyo

Andreas Pomp

The Boundary-Domain Integral Method for Elliptic Systems

 Springer

Author

Andreas Pomp
ISE Integrated Systems Engineering AG
Technoparkstrasse 1
CH-8005 Zürich, Switzerland
e-mail: pomp@ise.ch

Formerly:
University of Stuttgart
Mathematical Institute A, Germany

Cataloging-in-Publication Data applied for

Die Deutsche Bibliothek - CIP-Einheitsaufnahme

Pomp, Andreas:
The boundary domain integral method for elliptic systems / Andreas
Pomp. - Berlin ; Heidelberg ; New York ; Barcelona ; Budapest ;
Hong Kong ; London ; Milan ; Paris ; Santa Clara ; Singapore ;
Tokyo : Springer, 1998
  (Lecture notes in mathematics ; 1683)
  ISBN 3-540-64163-7

Mathematics Subject Classification (1991):
Primary: 65-02
Secondary: 35A08, 45-02, 46F99, 65N38, 65N99, 65R20, 73V10, 73V20

ISSN 0075-8434
ISBN 3-540-64163-7 Springer-Verlag Berlin Heidelberg New York

© Springer-Verlag Berlin Heidelberg 1998
Printed in Germany

Typesetting: Camera-ready T$_E$X output by the author
SPIN: 10651049   46/3142-543210 - Printed on acid-free paper

Table of Contents

Part II. Application to the Shell Model of Donnell-Vlasov-Type

Abbreviations and Symbols

BEM	boundary element method
FEM	finite element method
BDIM	boundary-domain integral method
DV model	shell model of Donnell-Vlasov-type

$\longmapsto$	a continuous mapping
$\hookrightarrow$	a compact mapping
$\mathfrak{Re}, \mathfrak{Im}$	real part and imaginary part

Sets and Spaces

$\mathbb{N}_0$	natural numbers $\{0, 1, 2, \ldots\}$
$\mathbb{N}$	positive natural numbers $\{1, 2, 3, \ldots\}$
$\mathbb{Z}$	integers
$\mathbb{R}^n$	n-tupel of real numbers ($n \in \mathbb{N}$)
$\mathbb{C}^n$	n-tupel of complex numbers ($n \in \mathbb{N}$)
$\mathbb{R}_+$	positive real numbers
S_n	unit sphere in $\mathbb{R}^n$
$C_0^\infty(\mathbb{R}^n)$	infinitely differentiable functions with compact support in $\mathbb{R}^n$
$C_0^\infty(\mathbb{R}^n \setminus 0)$	infinitely differentiable functions with compact support in $\mathbb{R}^n \setminus 0$
$\mathcal{S}(\mathbb{R}^n)$	rapidly decreasing C^∞ functions
$\mathcal{D}'$	all distributions = linear continuous functionals on $C_0^\infty(\mathbb{R}^n)$
$\mathcal{S}'$	tempered distributions = linear continuous functionals on $\mathcal{S}(\mathbb{R}^n)$
$C_\infty^{(\mu)}$	infinitely differentiable functions on $\mathbb{R}^n \setminus 0$, homogeneous of degree μ
$Hom^{(\mu)}$	homogeneous polynomials of degree μ

Tensor notation	**(only in part II) B**
$\alpha, \beta, \gamma, \ldots$	greek indizes take the values 1 and 2
$j, k, l, \ldots$	latin indizes take the values 1, 2, 3
summation convention	If an index appears twice (once as upper and once as lower) then the sum is taken from 1 to 2 or 1 to 3 for greek or latin indices, respectively.

Introduction

The boundary-domain integral method (BDIM) is a numerical method for the approximate solution of partial differential equations. It is also known under other names, e. g., as "parametrix method". The BDIM can be considered as a generalization of the well-known boundary element method (BEM), but a better classification seems to be if both terms, BEM and BDIM, are considered as different methods which fall under the generic term "integral equation methods".

Now, the finite element method (FEM) is the most general and the most used numerical method for solving partial differential equations. Why do we need other methods? This question has already provoked many discussions.

From many fields of mathematical physics, especially from computational mechanics, it is known that the BEM represents (beside FEM) a second powerful method for the numerical solution of partial differential equations (see e. g. [13]) and that each method has his specific advantages. If both methods are available, these advantages can be better employed by combined (possibly coupled) use.

The initial aim of our investigations was an application of the boundary element method (BEM) to the shell equations. It is well-known that the BEM can be applied only if a *fundamental solution* is available. But this is the critical point and the construction of fundamental solutions for shell equations is quite difficult. This was pointed out already several times in the literature (see e. g. [57, sect. 1.17]). For that reason, the FEM for shells is much more developed than the BEM and a lot of literature exists in this field.

The main difficulty results from the fact that no applicable method is known to construct fundamental solutions for partial differential equations[1] with variable coefficients - except in some very special cases [19, 20]. All usual shell models, however, lead to equations with variable coefficients, unless one restricts themself to one of the following special cases:

1. the mid-surface is spherical or circular cylindrical;
2. the shell is considered as *shallow*.

---

[1] Only the case of linear equations is considered here.

A lot of literature is devoted to the BEM for shells or its foundation, namely the construction of fundamental solutions. The last topic is closely related to the investigation of point loads, acting on a shell. The first papers on this problem date back to about the middle of the 40's. All papers on this field known to us, make at least one of both assumtions from above. The papers [2, 3, 28, 41, 49, 56, 58, 70, 76, 78, 79, 80, 83, 91, 100, 101, 102, 108, 117, 120, 123, 136, 138] are devoted to circular cylindrical or sperical shells. Shells with an other special geometry (in the shallow case) are considered in [44] and [119]. Investigations of fundamental solutions and boundary element methods for shells with an arbitrary curved mid-surface can be found in [15, 16, 17, 32, 40, 42, 43, 47, 71, 84, 92, 94, 95, 96, 106, 107, 118, 124, 125, 126, 127, 128, 130, 132, 137]. But shallowness is always assumed in these papers.

In the literature, however, one finds different interpretations of this terminology. That is why we will first explain, what is to be understood here under the "shallowness" of a shell. This assumption says:

$$1 + \frac{\vartheta}{R} \cong 1. \qquad (0.1)$$

ϑ denotes the maximal deviation of the considered shell sector from an approximating tangential plane and R is the miminal absolute radius of curvature within this section (see Fig. 1.1). The sign "$\cong$" in (0.1) means that in the shell equations all terms of magnitude ϑ/R can be neglected compared to unity. Especially this applies to the Christoffel symbols so that all covariant derivatives can be replaced by partial ones.

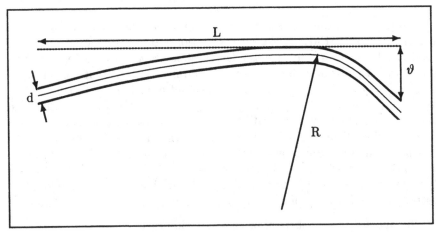

Fig. 1.1. A shell section with tangential plane

From elementary trigonometrical considerations (Pythagoras theorem), the following equivalent formulations of assumption (0.1) can be deduced:

$$1 + \left(\frac{L}{R}\right)^2 \cong 1 \qquad \text{or} \qquad 1 + \left(\frac{\vartheta}{L}\right)^2 \cong 1. \tag{0.2}$$

Clearly, the second criterion is applicable only in the non-oscillating case.

The shallowness assumption says that the considered shell sector has a diameter L so small, that changing values of the geometry-dependent functions, namely the components of the first and second fundamental tensor, can be neglected. In other words, this functions can be frozen at one fixed point without making an essential error. Then one obtains equations with constant coefficients if the shallowness assumption (0.1) is considered as fulfilled.

The thumb rule $\vartheta/L \leq 1/8$ is given in [100] for the question, when a shell can be considered as shallow for practical purposes.

If one is interested in an accurate calculation of a complex curved structure, it must be subdivided into a large number of sectors in order to fulfill the shallowness assumption (0.1) or (0.2) satisfactory well. Then, however, each one of these sectors will be not too much larger than the shell elements used in FEM. Therefore, an essential advantage of the BEM is lost. This was the motivation to develop an integral method which is also applicable to non-shallow shells.

An additional motivation origins from the fact that for shell equations with constant coefficients, the efficient calculation of the fundamental solution already appears as a serious problem. In the cases of a special mid-surface geometry, closed formulas for the fundamental solution can be derived, which contain special functions, namely Kelvin functions and modified Bessel functions. (A survey of these results can be found in [118].) In the case of arbitrary curved shells, only series expansions of the fundamental solution can be obtained anyway. The papers to this topic mostly use "Hörmander's cofactor method" and "plane wave decomposition" (cmp. sect. 2.10) for the derivation of the series representation. But the evaluation of the series for each pair of discretization points appears to be very expensive in practice.

An alternative consists in the *boundary-domain integral method* (BDIM), also known as *parametrix method*. It goes back to E. E. Levi [81] and D. Hilbert [59, 60]. Here, a boundary value problem is transformed into a coupled system of boundary and domain integral equations by using a *Levi function*. This is a function which contains the same singularity as the fundamental solution.

An application of the BDIM to the shell equations was already suggested by H. Antes [2]. In this paper, the method is explained for the case of circular cylindrical shells. D. E. Beskos [11], [12] used the BDIM for numerical investigations of shallow shells with an arbitrary curved mid-surface in the static and dynamic case. In [12, Introduction] one can find a remarkable assertion, based on numerical experiments, namely that the BDIM doesn't lead to higher computational time than the BEM, though the dimensions of the discretization matrices are essentially higher. This observation can be explained with the high expenses for evaluating the fundamental solution, as already mentioned.

This gives a further reason to develop the BDIM for shells. Since the BDIM can handle as well with variable coefficients, restrictions to shallow shells or shells with a special geometry can be dropped. For such problems (for shell equations with variable coefficients), to the best of our knowledge, an integral method is applied here for the first time.

The transformation into systems of integral equations was already widely used for analytical investgations of partial differential and pseudodifferential operators. Per exemple we mention only [52, chapter 5]. For numerical purposes, this method was also already used in some special cases, e. g., for wave propagation, heat flux or elasticity problems in inhomogeneous bodies (see [21, 121] and the papers referenced there).

All papers known to us, which apply the BDIM for numerical purposes (not only to solve the shell equations but also other problems), use only the main singularity for the Levi function. It seems to be of advantage and, as will be shown in section 2.5, sometimes even necessary to include further terms from the pseudohomogeneous series expansion of the fundamental solution into the Levi function. The Fredholm integral equation of the second kind, which is to be solved in the domain, then gets a smoother kernel and better properties. Likewise, the occurence of hypersingular integrals can avoided then be, at least if the Dirichlet problem for shells is considered.

Some question arise in this connection, namely:

1. How to calculate the higher order terms in the pseudohomogeneous series expansion of the fundamental solution?
2. Which consequences does the use of such "enhanced Levi functions" have for the solvability properties of the integral equations itself and the discretized systems?

Generally it appears that only very little has been done up to now [133, 134], concerning a mathematical analysis of the BDIM from the numerical point of view, though the basic ideas are very old and the BDIM was frequently applied for several practical problems. The aim of this book is to give a mathematical foundation of the BDIM from a numerical point of view and a detailed description of the complete algorithm.

The text is divided into two parts, each in several chapters. Each chapter contains a preamble with a summary of its contents. The aim of part I and the first main result is an algorithm to construct Levi functions of arbitrary degree[2] for elliptic systems of partial differential equations with variable coefficients. The general theory is applied in part II to a special shell model,

---

[2] The degree of a Levi function is introduced in section 2.2.

namely the model of the Donnell-Vlasov-type. The mid-surface is assumed to be smooth and arbitrarly curved. Its boundary may have corners. The pre-image in the parameter plane is a curvilinear polygon.

For this case, the construction of Levi functions is explained as well as the transformation of the shell equations into a system of integral equations. We will study the mapping and solvability properties of the particular integral operators and the complete system. Special emphasis is devoted to the singularities which occur in the vicinity of each corner point.

Further we will investigate a Galerkin method for numerical solution of the integral equation system. It uses piecewise constant trial functions in the domain and hat-functions on the boundary. If a graded meash is used, well-adapted to the corner singularities, a convergence order of (almost) h^2 can be achieved for the L^2-error of the displacement field. (h is the meshwidth.) This assertion is formulated in Theorem 6.10.1. A lot of preliminary theorems and lemmas are involved here so that Theorem 6.10.1 can be considered as the second main result.

The third main result consists in the observations from the numerical experiments, which lead to following conclusions:

- Besides the FEM we have with the BDIM a second universal, powerful, stable and efficient method for the numerical solution of shell equations, likewise, for elliptic equations (or systems) with variable coefficients.
- Due to the interiour degrees of freedom the BDIM produces discretization matrices with a high-order dimension, namely h^{-2}. But this dimension and the computational time keeps moderate if the accuracy of the solution is required inside realistic bounds (say, of about 2 or 3 valid digits).
- Contrary to the widespread opinion, a Galerkin method does not lead to essentially higher computational time than the collocation method, provided that not more quadrature sample points are used than necessary.

If one measures the computation time and the needed storage, the BDIM can compete with the FEM or appears even to be somewhat better. Likewise, one has no trouble with "locking-effects" and the BDIM is better adapted to corner singularities. A disadvantage, however, consists in the higher code developing time.

The algorithmic aspect stands in the foreground of the present book, as already mentioned. But many other mathematical topics come into consideration here, especially: shell theory, differential geometry, functional analysis, theory of distibutions, pseudodifferential operators, singular integral operators, approximation theory. These fields are handled here as far, as they are needed for the foundation of the BDIM.

Not investigated here, but of great importance in this connection is the solution of the discretized systems. Mutigrid and panel clustering, well-known techniques to increase the efficiency of boundary element methods (see [53] and other papers of W. Hackbusch), can be applied as well to the BDIM.

The treatement of Neumann or mixed problems is also not investigated here. In principle it is clear how the BDIM works in these cases. One has to deal then with the same difficulties as they are known from the BEM, especially for the evaluation of hypersingular integrals.

A planed application of the BDIM described here consists in the generation of the stiffness matrix for so called "macro shell-elements". Coupled with other macro elements or finite elements, this will be a powerful method to calculate stress distributions in shell structures. So, it is a basic tool for domain-decomposition and substructure techniques, what is again the base for efficient parallel algorithms.

The text starts from very theoretical things and comes step by step to more practical questions. The individual chapters are rather self-contained such that readers, which are more interested in the practical realization can start at a later point as well.

Acknowledgement. First of all I wish to thank Prof. W. Wendland for the initial suggestion to deal with this topic, for his permanent interest and many helpful discussions. Further I would like to thank Prof. G. Hsiao, Dr. S. Sauter, Prof. M. Costabel, Dr. J. Elschner and Dr. O. Steinbach for essential hints to special questions. I am also indebted to Mr. P. Richardson and Dr. R. Rühl for performing proof-readings of the manuscript.

A special thanks to the German Research Foundation, Bonn which supported this project during the period from 1990 to 1995 within SFB 259 "High Temperature Problems of Re-usable Space Transport Systems" and from 1995 to 1996 by a habilitation grant for the author.

Last, but foremost, I whish to dedicate this book to my wife as some small acknowledgement of her unfailing love and moral support.

Part I

The General Theory for Elliptic Systems of Partial Differential Equations

1. Pseudohomogeneous Distributions

This chapter has preliminary character. Some special classes of functions and distributions are introduced. The Levi-functions constructed in the next chapter will belong to them. Especially, the mapping properties of the basic operators

- multiplication with smooth functions,
- differentiation,
- right-inverse to homogeneous elliptic differential operators

between these classes will be of importance.

1.1 Basic Definitions of Homogeneity and Pseudohomogeneity

Denote by $\mathcal{D}'(\mathbb{R}^n)$ the set of all distributions on the space $C_0^\infty(\mathbb{R}^n)$ and by $\mathcal{S}'(\mathbb{R}^n)$ the subset of *tempered distributions*. A distribution $v \in \mathcal{D}'(\mathbb{R}^n)$ is called a *continuation* of the function $u(x) \in L^1_{\text{loc}}(\mathbb{R}^n \setminus 0)$ if

$$\langle v, \phi \rangle = \langle u, \phi \rangle, \qquad \forall \phi \in C_0^\infty(\mathbb{R}^n \setminus 0).$$

A function $u(x) \in L^1_{\text{loc}}(\mathbb{R}^n \setminus 0)$ is called *homogeneous of degree μ* if

$$u(\tau x) = \tau^\mu u(x), \qquad \forall \tau > 0, \forall x \neq 0.$$

Generally μ stands for an arbitrary complex number. Within the framework of this paper, however, only integer degrees are of interest. Thus we assume $\mu \in \mathbb{Z}$.

Sums of the form $E(x) + Q(x) \ln|x|$ with homogeneous functions[1] $E(x)$ and $Q(x)$ of degree μ are called *pseudohomogeneous functions of degree μ*. This terminology is adopted from [114]. The concept of "associated homogeneous functions" was introduced in [46, chapter I] for a slightly larger class which contains also higher powers of $\ln|x|$. Further we refer to the monograph [51] where the much more general concept of "quasihomogeneous distributions" is investigated into.

---

[1] Within the framework considered here, $Q(x)$ will always be a homogeneous polynomial.

The concept of homogeneity and pseudohomogeneity can be carried over in a natural manner to distributions. A distribution $v \in \mathcal{D}'(\mathbb{R}^n)$ is called a *homogeneous distribution of degree μ* if

$$\langle v, \phi \rangle = \tau^\mu \langle v, \phi_\tau \rangle, \qquad \forall \phi \in C_0^\infty(\mathbb{R}^n), \ \forall \tau > 0.$$

Here we have used the abbreviation $\phi_\tau(x) := \tau^n \phi(\tau x)$ (cmp. [62, Def. 3.2.2]). Similarly, a distribution $v \in \mathcal{D}'(\mathbb{R}^n)$ is called a *pseudohomogeneous distribution of degree μ* if there exists a homogeneous distribution $w \in \mathcal{D}'$ of degree μ such that

$$\langle v, \phi \rangle = \tau^\mu \left(\langle v, \phi_\tau \rangle + \ln \tau \langle w, \phi_\tau \rangle \right), \qquad \forall \phi \in C_0^\infty(\mathbb{R}^n), \ \forall \tau > 0.$$

We want to introduce some further notations. Let μ be an integer. The set of all functions $u(\xi) \in C^\infty(\mathbb{R}^n \setminus 0)$ homogeneous of degree μ is denoted by $C_\infty^{(\mu)}$. The subset of all homogeneous polynomials of degree μ is denoted by $\mathrm{Hom}^{(\mu)}$. For $\mu < 0$ we set $\mathrm{Hom}^{(\mu)} = \{0\}$.

Now we want to consider some special distributions in the one-dimensional case $n = 1$. The distributions t_+^μ defined for $\mu > -1$ by

$$\langle t_+^\mu, \phi(t) \rangle := \int_0^\infty t^\mu \phi(t) dt \tag{1.1}$$

are homogeneous distributions of degree μ. For $\mu = -k$, $(k \in \mathbb{N})$ there exist only pseudohomogeneous continuations of the homogeneous function $[(t + |t|)/2]^\mu$. Such a continuation (denoted by t_+^μ) is defined by the *partie-finie* limit (for $\varepsilon \to 0$)

$$
\begin{aligned}
\langle t_+^\mu, \phi(t) \rangle &:= \mathrm{p.f.} \int_\varepsilon^\infty t^\mu \phi(t) dt \\
&= \frac{1}{(k-1)!} \left[-\int_0^\infty \phi^{(k)}(t) \ln t \, dt + \phi^{(k-1)}(0) \sum_{j=1}^{k-1} \frac{1}{j} \right].
\end{aligned}
\tag{1.2}
$$

A definition of the partie-finie limit can be found e. g. in [62, sect. 3.2] or, in a slightly more general form (including also logarithmic terms) in [74, Def. 1.1.1] and [75]. The set of all pseudohomogeneous continuations of the function $[(t + |t|)/2]^{-k}$ is obtained if a multiple of the $(k-1)$th derivative of the δ-distribution is added to (1.2).

The distributions t_+^μ defined by (1.1) and (1.2) have the following property (cmp. [62, formula (3.2.1)']):

$$\langle t_+^\mu, t\phi(t) \rangle = \langle t_+^{\mu+1}, \phi(t) \rangle, \qquad \forall \phi \in C_0^\infty(\mathbb{R}^1), \ \forall \mu \in \mathbb{Z}. \tag{1.3}$$

1.2 Spherical Harmonics

The following notations and basic properties of spherical harmonics up to
formula (1.6) can be found in [86, chapter 14]. Note that the roles of m and
n are changed there. In our notation, n is the dimension of space and we
assume $n \geq 2$ within this section. Everything can be carried over to the case
$n = 1$, however, some of the formulas must then be modified.

Denote by $Y_{m,n}^{(k)}$ the n-dimensional spherical harmonics of m-th order ($m \in$
$\mathbb{N}_0$). The index k runs from 1 to

$$K_{m,n} := \begin{cases} (2m + n - 2)\frac{(n+m-3)!}{(n-2)!\,m!}, & n + m \geq 3, \\ 1, & m = 0,\ n = 2. \end{cases} \tag{1.4}$$

The n-dimensional harmonical polynomials of m th order form a linear vector
space of dimension (1.4) which is spanned by the harmonical polynomials
$r^m Y_{m,n}^{(k)}(\omega)$. Hereby, $(r; \omega)$ are the spherical coordinates of a point $x \in \mathbb{R}^n$.
The vector ω runs through the unit sphere S_n.

In spherical coordinates, Laplace's operator reads as

$$\Delta = \frac{\partial^2}{\partial r^2} + \frac{n - 1}{r}\frac{\partial}{\partial r} - \frac{1}{r^2}\Xi$$

with Beltrami's operator Ξ. The spherical harmonics are eigenfunctions of
Ξ:

$$\Xi\, Y_{m,n}^{(k)}(\omega) = m(m + n - 2)\, Y_{m,n}^{(k)}(\omega).$$

Hence

$$\Delta\left[r^p Y_{m,n}^{(k)}(\omega)\right] = (p - m)(p + m + n - 2)\, r^{p-2} Y_{m,n}^{(k)}(\omega), \qquad \forall p \in \mathbb{R}^1. \tag{1.5}$$

The spherical harmonics form an orthonormal system in the Hilbert space
$L^2(S_n)$, and thus each function $E(\omega) \in L^2(S_n)$ can be expanded into a Fourier
series

$$E(\omega) = \sum_{m=0}^{\infty} \sum_{k=1}^{K_{m,n}} E_m^{(k)} Y_{m,n}^{(k)}(\omega). \tag{1.6}$$

Lemma 1.2.1. Let $\mu \in \mathbb{N}_0$ and $E(\omega) \in L^2(S_n)$. Then the orthogonality
relation $\int_{S_n} E(\omega)\omega^\beta d\omega = 0$ is valid for all multi-indices β with $|\beta| = \mu$ if and
only if

$$E_{\mu-2j}^{(k)} = 0, \qquad \forall j = 0, \ldots, \left[\frac{\mu}{2}\right], \quad \forall k = 1, \ldots, K_{\mu-2j,n}.$$

Proof. We have to show that

$$\mathrm{span}\left\{\omega^\beta \,\Big|\, |\beta| = \mu\right\} = \tag{1.7}$$

$$= \mathrm{span}\left\{Y_{\mu-2j,n}^{(k)}(\omega) \,\Big|\, j = 0, \ldots, [\mu/2];\ k = 1, \ldots, K_{\mu-2j,n}\right\}.$$

First we remark that the inclusion ... $\supset$ ... in (1.7) is obvious, since for each pair of indices j, k from the specified range, the function $r^\mu Y^{(k)}_{\mu-2j,n}(\omega)$ is a homogeneous polynomial in x of degree μ. Hence, it can be written as a linear combination of polynomials x^β with $|\beta| = \mu$. It remains to show that the dimensions on both sides of (1.7) conincide.

The dimension of the left-hand side of (1.7) can be calculated by elementary combinatorial considerations as

$$\dim \operatorname{span} \left\{ \omega^\beta \Big| : |\beta| = \mu \right\} = \binom{\mu + n - 1}{\mu}. \tag{1.8}$$

The coincidence of this value with the sum $\sum_{j=0}^{[\mu/2]} K_{\mu-2j,n}$ can be verified by induction with respect to μ.   ∎

Lemma 1.2.2. *Let $\mu \in \mathbb{N}_0$ and let $v(x)$ be a homogeneous function of degree $\mu - 2$, the restriction of which to the unit sphere S_n is square integrable. Then Poisson's equation $\Delta u(x) = v(x)$ has a homogeneous solution $u(x)$ of degree μ in $\mathbb{R}^n \setminus 0$ if and only if the following Fourier coefficients of $v(x)$ vanish:*

$$v^{(k)}_\mu = 0, \qquad \forall k = 1, \ldots, K_{\mu,n}. \tag{1.9}$$

Proof. If a homogeneous solution exists, it has the series expansion

$$u(x) = r^\mu \sum_{m=0}^\infty \sum_{k=1}^{K_{m,n}} u^{(k)}_m Y^{(k)}_{m,n}(\omega), \tag{1.10}$$

convergent, due to classical regularity results, in each n-dimensional hollow sphere in the sense of L^2. By virtue of (1.5), the application of Laplace's operator gives

$$\Delta u(x) = r^{\mu-2} \sum_{m=0}^\infty \sum_{k=1}^{K_{m,n}} (\mu - m)(\mu + m + n - 2) u^{(k)}_m Y^{(k)}_{m,n}(\omega).$$

Hence, the Fourier coefficients

$$v^{(k)}_m = (\mu - m)(\mu + m + n - 2) u^{(k)}_m \tag{1.11}$$

vanish for $m = \mu$.

If, vice versa, condition (1.9) is fulfilled then the coefficients $u^{(k)}_m$ are determined from (1.11) for all $m \neq \mu$. The square summability of the sequence of numbers $v^{(k)}_m$ implies the square summability of $u^{(k)}_m$ since the later sequence decays faster for $m \to \infty$. With an arbitrary choice of the coefficients $u^{(k)}_\mu$, the series (1.10) represents then a homogeneous solution of $\Delta u = v$.   ∎

1.3 Fourier Transform

The Fourier transform of a test function $\phi \in C_0^\infty(\mathbb{R}^n)$ is defined as in [62] by

$$(\mathcal{F}\phi)(\xi) = \hat{\phi}(\xi) := \int e^{-i(x,\xi)}\phi(x)dx.$$

Accordingly, the Fourier transform $\hat{u}$ of a distribution $u \in S'$ is determined by $\langle \hat{u}, \phi \rangle = \langle u, \hat{\phi} \rangle$, $\forall \phi \in S(\mathbb{R}^n)$. With the abbreviation $D_j := -i\partial_j$, the following basic relations hold [62, formula (7.1.15)]:

$$\widehat{D_j u} = \xi_j \hat{u}, \quad \widehat{x_j u} = -D_j \hat{u}, \quad \forall u \in S'. \tag{1.12}$$

For later use we will collect some further important properties of the Fourier transform.

Lemma 1.3.1. [62, Theorem 7.1.10] *The Fourier transform is a bijective mapping from S' onto S' (in the topology of S').*

Lemma 1.3.2. [62, Theorem 7.1.16] *If $u \in S'$ is homogeneous of degree μ then $\hat{u}$ is homogeneous of degree $-(n + \mu)$. (This assertion holds for all complex numbers μ.)*

Lemma 1.3.3. [62, Theorem 7.1.18] *If $v \in \mathcal{D}'$ is the continuation of a homogeneous function u, then $v \in S'$. If, additionally, $u \in C^\infty(\mathbb{R}^n \setminus 0)$, then $\hat{v}$ is the continuation of a function from $C^\infty(\mathbb{R}^n \setminus 0)$.*

1.4 The Canonical Continuation of Homogeneous Functions

In the sequel we need the Operators R_μ defined by

$$(R_\mu \phi)(x) := \langle t_+^{n+\mu-1}, \phi(tx) \rangle, \quad \forall x \neq 0, \tag{1.13}$$

which map each test function $\phi \in C_0^\infty(\mathbb{R}^n)$ for $\mu > -n$ to a homogeneous function of degree $-(n+\mu)$ and for $\mu \leq -n$ to a pseudohomogeneous function of degree $-(n + \mu)$. These mapping properties will be shown later by means of formula (1.22). In the case $\mu = -n$, for instance, the operator R_μ has the representation

$$(R_{-n}\phi)(x) = -\int_0^\infty \frac{\partial\phi(tx)}{\partial t}\ln t\, dt = -\sum_{k=1}^n x_k \int_0^\infty \phi_{,k}(tx)\ln t\, dt. \tag{1.14}$$

This can be obtained from (1.2).

For later purposes we calculate the derivative of (1.13) with respect to x_j, whereby (1.3) is used:

$$\frac{\partial}{\partial x_j}\left(R_{\mu-1}\phi\right)(x) = \left\langle t_+^{n+\mu-2}, \frac{\partial}{\partial x_j}\phi(tx)\right\rangle = \left\langle t_+^{n+\mu-2}, t\phi_{,j}(tx)\right\rangle = \left(R_\mu\phi_{,j}\right)(x).$$
(1.15)

The commutativity of differentiation and integration can be deduced from the distribution property of t_+^μ or directly from (1.2).

Definition 1.4.1. *A distribution $u^\bullet$ defined by*

$$\langle u^\bullet, \phi\rangle \quad := \quad \int_{|\omega|=1} u(\omega)\left(R_\mu\phi\right)(\omega)d\omega, \qquad \forall \phi \in C_0^\infty\left(\mathbb{R}^n\right) \quad (1.16)$$

$$= \quad p.f.\int_{|x|>\varepsilon} u(x)\phi(x)dx \qquad\qquad\qquad (1.17)$$

is assigned to each homogeneous function $u(x)$ of degree μ. It is called the "canonical continuation" of u (cmp. [74]).

The coincidence of (1.16) and (1.17) is proved in [74, section 1.1]. In the special case that u is homogeneous of degree $-n$, we conclude by insertion of (1.14) into (1.16):

$$\langle u^\bullet, \phi\rangle \quad = \quad -\sum_{k=1}^n \int_{|\omega|=1}\int_0^\infty t^n u(t\omega)\omega_k\,\phi_{,k}(t\omega)\ln t\, dt d\omega$$

$$= \quad -\sum_{k=1}^n \int_{\mathbb{R}^n} u(x)x_k\phi_{,k}(x)\ln|x|\, dx.$$

These are convergent integrals since $x_k u(x)$ is homogeneous of degree $1-n$. Note that $dx = t^{n-1}dtd\omega$.

An other approach to define the continuation of a homogeneous function u of degree μ is given by [62, formula (3.2.23)]:

$$\langle u^\bullet, \phi\rangle := \langle u, \psi R_\mu\phi\rangle, \qquad \forall \phi \in C_0^\infty\left(\mathbb{R}^n\right). \qquad (1.18)$$

Hereby, ψ is an arbitrary test function from $C_0^\infty\left(\mathbb{R}^n \setminus 0\right)$ fulfilling the properties[2]:

$$\int_0^\infty \frac{\psi(\tau x)}{\tau}d\tau \quad = \quad 1, \qquad \forall x \text{ with } |x|=1, \qquad (1.19)$$

$$\int_0^\infty \frac{\psi(\tau x)}{\tau}\ln\tau\, d\tau \quad = \quad 0, \qquad \forall x \text{ with } |x|=1. \qquad (1.20)$$

The existence of such test functions can be verified easily. They can be choosen rotationally symmetric.

---

[2] Only the property (1.19) is required in [62]. This explains the remark after the proof of [62, Theorem 3.2.3] which says that the functionals of the form (1.18) may depend on ψ in the case $\mu \le -n$.

Lemma 1.4.1. *The distibution $u^\bullet$ defined by (1.18) does not depend on ψ and coincides with the canonical continuation (1.16).*

Proof. For $k \in \mathbb{N}_0$ we denote by P_k the projection which takes from the Taylor expansion of a sufficiently smooth function $\phi(x)$ only the homogeneous part of degree k:

$$(P_k\phi)(x) := \sum_{|\beta|=k} \frac{\phi^{(\beta)}(0)}{\beta!} x^\beta = \left[\frac{1}{k!}\frac{d^k}{d\tau^k}\phi(\tau x)\right]\Bigg|_{\tau=0}. \qquad (1.21)$$

Further we set $P_k = 0$ for $k < 0$.

The homogeneity of the functionals (1.1) (cmp. [62, formula (3.2.7)]) and the pseudohomogeneity of the functionals (1.2) (cmp. [62, formula (3.2.8)]) can be written now in compact form:

$$\langle t_+^{-k}, \phi_\tau(t)\rangle = \tau^k \left[\langle t_+^{-k}, \phi(t)\rangle - \ln\tau (P_{k-1}\phi)(0)\right], \qquad \forall k \in \mathbb{Z}.$$

Hence we have

$$\left\langle t_+^{-(k+1)}, \phi(tx)\right\rangle = \tau^{-k}\left\langle t_+^{-(k+1)}, \phi(\tau tx)\right\rangle + \ln\tau \,(P_k\phi)(x), \qquad \forall k \in \mathbb{Z},$$

$$\begin{aligned}(R_\mu\phi)(\tau x) &= \left\langle t_+^{n+\mu-1}, \phi(\tau tx)\right\rangle \qquad\qquad\qquad (1.22)\\ &= \tau^{-(n+\mu)}\left[(R_\mu\phi)(x) - \ln\tau\,\left(P_{-(n+\mu)}\phi\right)(x)\right], \quad \forall\mu \in \mathbb{Z}.\end{aligned}$$

This is valid for all $\tau > 0$, $x \neq 0$, $\phi \in C_0^\infty(\mathbb{R}^n)$. For a homogeneous function u of degree μ, the functional (1.18) can be written now in the form

$$\begin{aligned}\langle u, \psi R_\mu\phi\rangle &= \int_0^\infty \int_{|\omega|=1} \tau^{n-1} u(\tau\omega)\psi(\tau\omega)(R_\mu\phi)(\tau\omega)d\omega d\tau\\ &= \int_0^\infty \int_{|\omega|=1} u(\omega)\frac{\psi(\tau\omega)}{\tau}\left[(R_\mu\phi)(\omega) - \ln\tau\,\left(P_{-(n+\mu)}\phi\right)(\omega)\right]d\omega d\tau\\ &= \int_{|\omega|=1} u(\omega)(R_\mu\phi)(\omega)d\omega.\end{aligned}$$

The last expression coincides with (1.16). ∎

1.5 Properties of the Canonical Continuation

Let $u(x)$ be a homogeneous function of (integer) degree μ. Then:

1. $u^\bullet \in \mathcal{S}'(\mathbb{R}^n)$ (see Lemma 1.3.3).
2. For $\mu > -n$ the canonical continuation is the unique homogeneous continuation of u (see [62, Theorem 3.2.3]).

3. For $\mu = -(n+k)$ and $k \in \mathbb{N}_0$ there exist in general only pseudohomogeneous continuations of u. All these continuations have the representation

$$u^{\bullet} + \sum_{|\beta|=k} c_{\beta} \delta^{(\beta)}, \qquad (1.23)$$

i. e., they form a linear mannifold of dimension $\binom{k+n-1}{k}$. Hereby β denotes a multi-index, c_{β} are certain constants and $\delta^{(\beta)} = \partial^{\beta} \delta$ are derivatives of the δ-distribution. If and only if the $\binom{k+n-1}{k}$ additional conditions

$$\int_{|\omega|=1} u(\omega)\omega^{\beta} d\omega = 0, \qquad \forall |\beta| = k \qquad (1.24)$$

are fulfilled, then a homogeneous continuation exists. All continuations (1.23) are homogeneous in this case (cmp. [62, Theorem 3.2.4]).

4. For each homogeneous polynomial $H(x)$, there holds

$$[H(x)u(x)]^{\bullet} = H(x)u^{\bullet}. \qquad (1.25)$$

Such a commutator property is in general not valid for other continuations than the canonical one.

5. The canonical continuation does, in general, not commute with the derivatives $\partial_j = \frac{\partial}{\partial x_j}$. If u is homogeneous of degree $1 - n - k$ then:

$$\partial_j u^{\bullet} - (\partial_j u)^{\bullet} = \begin{cases} 0, & \text{for } k < 0, \\ (-1)^k \sum_{|\beta|=k} \frac{\delta^{(\beta)}}{\beta!} \int_{|\omega|=1} u(\omega)\omega_j \, \omega^{\beta} \, d\omega, & \text{for } k \geq 0. \end{cases} \qquad (1.26)$$

Here, the derivative $\partial_j u$ is understood in the sense of $\mathcal{D}'(\mathbb{R}^n \setminus 0)$.

Proof (of property 4). Let $H(x)$ be homogeneous of degree ν. From (1.3) we obtain

$$\left\langle t_+^{n+\mu-1}, H(tx)\phi(tx) \right\rangle = H(x) \left\langle t_+^{n+\mu+\nu-1}, \phi(tx) \right\rangle,$$

i. e., $R_{\mu}(H\phi) = H R_{\mu+\nu}\phi$. Now we apply (1.16):

$$\begin{aligned} \langle (Hu)^{\bullet}, \phi \rangle &= \langle Hu, \psi R_{\mu+\nu}\phi \rangle = \langle u, \psi H R_{\mu+\nu}\phi \rangle \\ &= \langle u, \psi R_{\mu}(H\phi) \rangle = \langle u^{\bullet}, H\phi \rangle = \langle Hu^{\bullet}, \phi \rangle. \end{aligned}$$

If we replace in (1.25) the canonical continuation by some other one, then the difference of both sides contains point functionals at most. The property (1.25) may be violated then, however, only for $\mu + \nu \leq -n$. ∎

Proof (of property 5). Let $\psi \in C_0^{\infty}(\mathbb{R}^n \setminus 0)$ be a test function fulfilling (1.19) and (1.20). If the property (1.19) holds for $|x| = 1$, then it holds for all $x \neq 0$. Thus,

$$0 = \partial_j \int_0^{\infty} \frac{\psi(\tau x)}{\tau} d\tau = \int_0^{\infty} \psi_{,j}(\tau x) d\tau.$$

Further, from (1.20) we obtain by use of the substitution $s = \tau|x|$:

$$\int_0^\infty \frac{\psi(\tau x)}{\tau} \ln \tau \, d\tau = \int_0^\infty \frac{1}{s} \psi\left(s \frac{x}{|x|}\right) \ln \frac{s}{|x|} ds = -\ln|x|,$$

$$\partial_j \int_0^\infty \frac{\psi(\tau x)}{\tau} \ln \tau \, d\tau = \int_0^\infty \psi_{,j}(\tau x) \ln \tau \, d\tau = -\frac{x_j}{|x|^2}.$$

Due to (1.13) and (1.22), this gives (with $k = 1 - n - \mu$):

$$[R_\mu (\psi_{,j} R_{\mu-1} \phi)] (x) = \int_0^\infty t^{n+\mu-1} \psi_{,j}(tx) (R_{\mu-1}\phi)(tx) dt$$

$$= (R_{\mu-1}\phi)(x) \int_0^\infty \psi_{,j}(tx) dt - (P_k\phi)(x) \int_0^\infty \psi_{,j}(tx) \ln t \, dt$$

$$= \frac{x_j}{|x|^2} (P_k\phi)(x).$$

Inserting $\psi_{,j} R_{\mu-1} \phi \in C_0^\infty (\mathbb{R}^n \setminus 0)$ instead of ϕ into (1.16) leads to:

$$\langle u, \psi_{,j} R_{\mu-1} \phi \rangle = \int_{|\omega|=1} u(\omega) [R_\mu (\psi_{,j} R_{\mu-1} \phi)] (\omega) d\omega$$

$$= \int_{|\omega|=1} u(\omega) \omega_j (P_k\phi)(\omega) \, d\omega$$

$$= \sum_{|\beta|=k} \frac{\phi^{(\beta)}(0)}{\beta!} \int_{|\omega|=1} u(\omega) \omega_j \omega^\beta d\omega.$$

It can be verified by use of (1.15) that the last expression coincides with

$$\langle \partial_j u^\bullet, \phi \rangle - \langle (\partial_j u)^\bullet, \phi \rangle = -\langle u^\bullet, \phi_{,j} \rangle - \langle \partial_j u, \psi R_{\mu-1} \phi \rangle$$

$$= -\langle u, \psi R_\mu \phi_{,j} \rangle + \langle u, \partial_j (\psi R_{\mu-1} \phi) \rangle = \langle u, \psi_{,j} R_{\mu-1} \phi \rangle.$$

Formula (1.26) is proved. ∎

The commutator formula (1.26) was already found by F. G. Tricomi for the special case $n = 2, k = 0$, cmp. [87, §1, formula(2c)].

1.6 The Canonical Projection

The canonical continuation can be considered as a rule[3] to select a special element from the linear manifold (1.23) of all pseudohomogeneous continuations of a homogeneous function. This selection rule can be associated with a projection $\widehat{T}$ defined by

---

[3] We remark that other possibilities for such a selection rule exist as well, e. g., the rule given by [35, formula (2.84)].

$$\widehat{\mathcal{T}} v := \left[v \big|_{\mathbb{R}^n \setminus 0} \right]^{\bullet}.$$

The domain of definition is the set of all distributions $v \in \mathcal{S}'$ which are continuations of a homogeneous function. On the Fourier image (= Fourier pre-image) of this set we can define the *"canonical projection"* $\mathcal{T}$ as follows:

$$\mathcal{T}u := \mathcal{F}^{-1}\widehat{\mathcal{T}}\mathcal{F}u.$$

Remark 1.6.1. If $u \in \mathcal{S}'$ is homogeneous of degree $\mu < 0$ then $\mathcal{T}u = u$.

This can be deduced from property 2 of the canonical continuation and Lemma 1.3.2.

Lemma 1.6.1. *Suppose $u(x) \in C_\infty^{(\mu)}$ with $\mu \in \mathbb{N}_0$. Then the commutator property*

$$\left(\mathcal{T}|x|^2 - |x|^2\mathcal{T} \right) v(x) = 0 \tag{1.27}$$

is valid for the distribution $v(x) = \Delta u(x)$.

Proof. Due to (1.12), the equation $\Delta u = v$ implies the relation $\widehat{v}(\xi) = -|\xi|^2\widehat{u}(\xi)$. Since $\widehat{u}$ is a homogeneous distribution of degree $-(n + \mu)$ (see Lemma 1.3.2), the conditions (1.24) are satisfied:

$$\int \widehat{u}(\omega)\omega^\beta d\omega = \int_{|\omega|=1} \widehat{v}(\omega)\omega^\beta d\omega = 0, \qquad \forall |\beta| = \mu.$$

The distribution $\widehat{v}$ is homogeneous of degree $2 - n - \mu$. Formula (1.26) leads to:

$$\left(\partial_j\widehat{\mathcal{T}} - \widehat{\mathcal{T}}\partial_j\right)\widehat{v} = (-1)^{\mu-1} \sum_{|\beta|=\mu-1} \frac{\delta^{(\beta)}}{\beta!} \int_{|\omega|=1} \widehat{v}(\omega)\omega_j\omega^\beta d\omega = 0,$$

$$\sum_{j=1}^n \widehat{v}_{,j}(\omega)\omega_j = \left.\frac{\partial\widehat{v}(t\omega)}{\partial t}\right|_{t=1} = (2 - n - \mu)\widehat{v}(\omega),$$

$$\sum_{j=1}^n \left(\partial_j\widehat{\mathcal{T}} - \widehat{\mathcal{T}}\partial_j\right)\widehat{v}_{,j} = (-1)^\mu \sum_{|\beta|=\mu} \frac{\delta^{(\beta)}}{\beta!} \int_{|\omega|=1} \left(\sum_{j=1}^n \widehat{v}_{,j}(\omega)\omega_j\right) \omega^\beta d\omega = 0,$$

$$\left(\Delta\widehat{\mathcal{T}} - \widehat{\mathcal{T}}\Delta\right)\widehat{v} = \sum_{j=1}^n \left[\partial_j\left(\partial_j\widehat{\mathcal{T}} - \widehat{\mathcal{T}}\partial_j\right)\widehat{v} + \left(\partial_j\widehat{\mathcal{T}} - \widehat{\mathcal{T}}\partial_j\right)\widehat{v}_{,j}\right] = 0.$$

From the last relation we obtain

$$\mathcal{F}^{-1}\left(\Delta\widehat{\mathcal{T}} - \widehat{\mathcal{T}}\Delta\right)\widehat{v} = \left(\mathcal{T}|x|^2 - |x|^2\mathcal{T}\right)v(x) = 0,$$

q. e. d.   ∎

The next theorem gives an insight, how the canonical continuation acts on the Fourier image. We obtain a duality relation to (1.24).

Theorem 1.6.1. *Let $E(x) \in C_\infty^{(\mu)}$ with $\mu \in \mathbb{N}_0$. Then*

$$(I - \mathcal{T})E(x) = 0 \quad \Longleftrightarrow \quad \int_{|\omega|=1} E(\omega)\omega^\beta d\omega = 0, \quad \forall |\beta| = \mu. \quad (1.28)$$

Proof. Denote by $E_\mu^{r/l}$ the linear vector space of all functions $E(x) \in C_\infty^{(\mu)}$ fulfilling the conditions on the right-/left-hand side of (1.28), respectively. At first we show the inclusion $E_\mu^r \subset E_\mu^l$.

Therefore, let $E(x)$ fulfill the property on the right-hand side. Due to Lemma 1.2.1, the coefficients of the Fourier expansion (1.6) satisfy the relations

$$E_{\mu-2j}^{(k)} = 0, \quad \forall j = 0, \ldots, \nu := \left[\frac{\mu}{2}\right], \quad \forall k = 1, \ldots, K_{\mu-2j,n}. \quad (1.29)$$

For each $j = 0, \ldots, \nu$ the function $v_j(x) := r^{-2(j+1)}E(x)$ is homogeneous of degree $\mu - 2(j+1)$ and the conditions of Lemma 1.2.2 are fulfilled. Hence, for all $j = 0, \ldots, \nu$ exists a homogeneous function $u_j(x)$ of degree $\mu - 2j$ which solves the equation $\Delta u_j = v_j$ in $\mathbb{R}^n \setminus 0$. Identifying u_j and v_j with its canonical continuations, equation $\Delta u_j = v_j$ holds in the distributional sense everywhere in $\mathbb{R}^n$.

The continuations v_j are also homogeneous. This can be verified in the special case that $n = 2$ and that v_ν is homogeneous of degree -2 using the fact that (1.29) implies (1.24). In all other cases the degree of homogeneity of v_j is greater than $-n$ and by virtue of property 2 the canonical continuation is homogeneous too.

Application of Lemma 1.6.1 gives:

$$\mathcal{T}\left(r^2 v_j\right) = r^2 \left(\mathcal{T}v_j\right), \quad \forall j = 0, \ldots, \nu. \quad (1.30)$$

Now, v_ν is homogeneous of degree -2 for even μ and homogeneous of degree -1 for odd μ. Due to Remark 1.6.1, there holds $\mathcal{T}v_\nu = v_\nu$. Inserting this into (1.30) gives:

$$\mathcal{T}v_{\nu-1} = \mathcal{T}\left(r^2 v_\nu\right) = r^2 v_\nu = v_{\nu-1}.$$

Repeated insertion into (1.30) leads to the relations $\mathcal{T}(r^2 v_j) = r^2 v_j, \forall j = 0, \ldots, \nu$. In the case $j = 0$ this is equivalent to the left-hand side of (1.28).

The inclusion $E_\mu^r \subset E_\mu^l$ involves the inequality codim $E_\mu^l \leq$ codim $E_\mu^r = \binom{\mu+n-1}{\mu}$. On the other hand, the Fourier transform of each homogeneous polynomial is a linear combination of derivatives of the δ-distribution. Thus $\mathcal{T}H(x) = 0$ holds for all polynomials $H(x)$, especially for the $\binom{\mu+n-1}{\mu}$ linear independent homogeneous polynomials x^β of degree $|\beta| = \mu$. This gives the opposite inequality codim $E_\mu^l \geq \binom{\mu+n-1}{\mu}$. Therefore here we have the sign of equality and the vector spaces E_μ^l and E_μ^r coincide since they have the same codimension. ∎

We still define the numbers

$$A^\gamma := \frac{1}{|S_n|} \int_{|\omega|=1} \omega^\gamma d\omega$$

for each multi-index $\gamma = (\gamma_1, \ldots, \gamma_n)$. If, at least, one of the indices $\gamma_1, \ldots, \gamma_n$ is odd then A^γ is equal to zero. Otherwise, if all indices are even, then

$$A^\gamma = \frac{(n-2)!!}{(|\gamma|+n-2)!!} \prod_{j=1}^{n} (\gamma_j - 1)!!. \tag{1.31}$$

This formula can be deduced from the explicit representation of the spherical coordinates and the integration rules for trigonometric polynomials.

Corollary 1.6.1. *The canonical projection applied to an arbitrary function* $E(x) \in C_\infty^{(\mu)}$ *gives*

$$(\mathcal{T}E)(x) = E(x) - U(x), \tag{1.32}$$

where $U(x) = \sum_{|\alpha|=\mu} U_\alpha x^\alpha \in Hom^{(\mu)}$ *is a homogeneuos polynomial, its coefficients* U_α *are the solution of the linear system*

$$\sum_{|\alpha|=\mu} A^{\alpha+\beta} U_\alpha = \frac{1}{|S_n|} \int_{|\omega|=1} E(\omega) \omega^\beta d\omega, \qquad \forall |\beta| = \mu. \tag{1.33}$$

The matrix of the system (1.33) *is symmetric and positive definite.*

The analogy to Theorem 1.6.1 in the case $n = 1$ reads as follows. If $E(x)$ is homogeneous of degree μ on both half-axes (not necessarily smooth at the origin) then

$$(I - \mathcal{T})E(x) = 0 \quad \Longleftrightarrow \quad E(1) + (-1)^\mu E(-1) = 0.$$

Formula (1.32) takes the following form for $n = 1$:

$$\begin{aligned} (\mathcal{T}E)(x) &= E(x) - \frac{1}{2} [E(1) + (-1)^\mu E(-1)] x^\mu \\ &= \frac{1}{2} [E(1) - (-1)^\mu E(-1)] x^\mu \operatorname{sgn}(x). \end{aligned}$$

1.7 The Canonical Projection Applied to Pseudohomogeneous Functions

The linear operator B is defined on the set of all homogeneous polynomials in $\mathbb{R}^n$ as follows. For each homogeneous polynomial $Q(x)$, the expression

$$\left[Q(\partial)\hat{\mathcal{T}} - \hat{\mathcal{T}}Q(\partial) \right] |\xi|^{-n} =: -BQ(\partial)\delta \tag{1.34}$$

can be calculated by use of the commutator formula (1.26). Expression (1.34) is a linear combination of derivatives of the δ-distribution. By replacing on the right-hand side $BQ(\partial)$ by $BQ(x)$ we receive those polynomial, which is defined as the image of Q by virtue of B. One can see that BQ has the same degree of homogeneity as Q.

For an explicit evaluation of the operator B we need the coefficients (1.31). For instance, the following mapping relations can be deduced:

$$\text{for} \quad Q(x) = \text{const.} \quad \text{is} \quad BQ = 0,$$

$$\text{for} \quad Q(x) = x_j \quad \text{is} \quad (BQ)(x) = \frac{|S_n|}{n}x_j,$$

$$\text{for} \quad Q(x) = x_j^2 \quad \text{is} \quad (BQ)(x) = \frac{|S_n|}{n+2}\left[\frac{2(n+1)}{n}x_j^2 + \frac{1}{2}r^2\right]$$

and so on. The unit speres have the measure [86, Kap. 1, §2, formula(9)]:

$$|S_n| = \frac{2\,\pi^{n/2}}{\Gamma\left(\frac{n}{2}\right)}.$$

The Fourier transform of the logarithmic function is calculated in [113, formula (VII, 7; 16)]. Taking into account that the Fourier transform is introduced there with other constants, we obtain after an appropriate conversion:

$$\mathcal{F}_{x\to\xi}\ln|x| = C_n\left[|\xi|^{-n}\right]^{\bullet} + C_n'\delta \tag{1.35}$$

with the constants

$$C_n := -2^{n-1}\pi^{n/2}\Gamma\left(\frac{n}{2}\right),$$

$$C_n' := \frac{(2\pi)^n}{2}\left[\frac{\Gamma'\left(\frac{n}{2}\right)}{\Gamma\left(\frac{n}{2}\right)} - C_E - 2\ln\pi\right].$$

Hereby, $C_E = 0.577215\ldots$ denotes Euler's constant.

Theorem 1.7.1. *Let $Q(x)$ and $R(x)$ be two homogeneous polynomials of the same degree in $\mathbb{R}^n$. Then there holds*

$$(I - \mathcal{T})[Q(x)\ln|x| - R(x)] = 0$$

if and only if

$$R(x) = \frac{1}{(2\pi)^n}[C_n'Q(x) - C_n(BQ)(x)]. \tag{1.36}$$

Proof.

$$\mathcal{F}[Q(x)\ln|x| - R(x)] = Q(-D)\mathcal{F}\ln|x| - R(-D)\mathcal{F}(1)$$

$$= Q(-D)\left\{C_n\left[|\xi|^{-n}\right]^{\bullet} + C_n'\delta\right\} - (2\pi)^nR(-D)\delta$$

$$= C_n\left[Q(-D)|\xi|^{-n}\right]^{\bullet} + C_n\left[Q(-D)\widehat{\mathcal{T}} - \widehat{\mathcal{T}}Q(-D)\right]|\xi|^{-n} +$$

$$+ C_n'Q(-D)\delta - (2\pi)^nR(-D)\delta.$$

If we apply $I - \widehat{T}$ to the last expression, the first summand vanishes obviously. The residual terms vanish if and only if

$$(2\pi)^n R(\partial)\delta = C'_n Q(\partial)\delta + C_n \left[Q(\partial)\widehat{T} - \widehat{T}Q(\partial) \right] |\xi|^{-n}.$$

This is equivalent to condition (1.36).   ∎

1.8 Special Classes of Pseudohomogeneous Distributions

The class $\Pi(\mu, 0)$ is defined as the set of inverse Fourier transforms $\mathcal{F}^{-1}(v)$ of all pseudohomogeneous continuations v of the form (1.23), where u traces $C_\infty^{(-n-\mu)}$. If we do not take all continuations, but only the canonical ones, we get the subclasses

$$\Pi^\bullet(\mu, 0) := \left\{ \mathcal{F}^{-1}(u^\bullet) \,\middle|\, : u(\xi) \in C_\infty^{(-n-\mu)} \right\}. \tag{1.37}$$

The inverse Fourier transform $\mathcal{F}^{-1}$ stands in both definitions for technical reasons. Replacing it by the Fourier transform $\mathcal{F}$ would not change the obtained classes.

By Lemma 1.3.1 and the property 1 of the canonical continuation follows immediately:

$$\Pi^\bullet(\mu, 0) \subset \Pi(\mu, 0) \subset \mathcal{S}'(\mathbb{R}^n).$$

Both, $\Pi^\bullet(\mu, 0)$ and $\Pi(\mu, 0)$ are infinite-dimensional linear vector spaces and the canonical projection has the mapping property

$$\mathcal{T}: \; \Pi(\mu, 0) \longmapsto \Pi^\bullet(\mu, 0). \tag{1.38}$$

Equivalently, the class $\Pi^\bullet(\mu, 0)$ could be defined as the image of $\mathcal{T}$ applied to $\Pi(\mu, 0)$.

Theorem 1.8.1. *For $\mu > -n$, the classes $\Pi(\mu, 0)$ and $\Pi^\bullet(\mu, 0)$ contain functions, else distributions. These functions or distributions have the following form:*

(i)   *For $\mu \geq 0$ there holds*

$$\Pi(\mu, 0) = \left\{ E(x) + Q(x) \ln |x| \,\middle|\, : E(x) \in C_\infty^{(\mu)}, \; Q(x) \in Hom^{(\mu)} \right\}. \tag{1.39}$$

A function $E(x) + Q(x) \ln |x| \in \Pi(\mu, 0)$ belongs to the subspace $\Pi^\bullet(\mu, 0)$ if and only if

$$(I - \mathcal{T})[E(x) + Q(x) \ln |x|] = (I - \mathcal{T})E(x) + R(x) = 0, \tag{1.40}$$

where $R(x)$ is derived from $Q(x)$ by virtue of (1.36).

(ii) For $-n < \mu < 0$ there holds

$$\Pi(\mu,0) = \Pi^\bullet(\mu,0) = C_\infty^{(\mu)}. \tag{1.41}$$

(iii) For $\mu \leq -n$, the spaces $\Pi(\mu,0)$ and $\Pi^\bullet(\mu,0)$ coincide. They consist of all distributions of the form

$$[E(x)]^\bullet + \sum_{|\beta|=-(n+\mu)} c_\beta \delta^{(\beta)} \tag{1.42}$$

with arbitrary constants $c_\beta \in \mathbb{C}$ and a function $E(x) \in C_\infty^{(\mu)}$ fulfilling the side conditions

$$\int_{|\omega|=1} E(\omega)\omega^\beta d\omega = 0, \qquad \forall |\beta| = -(n+\mu). \tag{1.43}$$

Proof.
1. The case $\mu < 0$: From $u \in \Pi(\mu,0)$ it follows by definition that the Fourier transform $\hat{u}$ is the pseudohomogeneuos continuation of a homogeneous function of degree $-(n+\mu)$. Property 2 from section 1.5 says that $\hat{u}$ consequently must be a canonical continuation, homogeneous of degree $-(n+\mu)$. This implies firstly the coincidence of $\Pi(\mu,0)$ with $\Pi^\bullet(\mu,0)$ and secondly (due to Lemma 1.3.2) that u itself is a homogeneous distribution of degree μ.

Further, $u \in \Pi(\mu,0)$ means that the restriction of $\hat{u}$ to $\mathbb{R}^n \setminus 0$ is a homogeneous C^∞ function. Lemma 1.3.3 implies that the Fourier transform of $\hat{u}$ is a C^∞ function if restricted to $\mathbb{R}^n \setminus 0$. This assertion holds in the same way for the inverse Fourier transform, i. e., for u.

Thus $u \in \Pi^\bullet(\mu,0)$ implies firstly that u is a homogeneous distribution of degree μ and secondly that u is the continuation of a function from $C_\infty^{(\mu)}$. From properties 2 and 3 of section 1.5 follows then immediately that u must have the form (1.41) or (1.42), respectively. The homogeneity requires in case $\mu \leq -n$ that the conditions (1.24) are satisfied.

Vice versa, if u is a distribution of the form (1.41) or (1.42) fulfilling the conditions (1.43) for $\mu \leq -n$, then u is a homogeneous distribution of degree μ. Lemma 1.3.2 and Lemma 1.3.3 imply that $\hat{u}$ is then the (unique) homogeneous continuation of a function from $C_\infty^{(-n-\mu)}$.

2. The case $\mu \geq 0$: A function $u \in \Pi^\bullet(\mu,0)$ must have the form (1.39). This follows immediately from [62, formula (7.1.19)]. Again, the C^∞ property of $E(x)$ for $x \neq 0$ is a consequence of Lemma 1.3.3. The assertion for $u \in \Pi(\mu,0)$ is obtained from the decomposition

$$\Pi(\mu,0) := \Pi^\bullet(\mu,0) + \text{Hom}^{(\mu)}, \tag{1.44}$$

which can be deduced from (1.23) and the properties of the Fourier transform.

Vice versa, let $u(x)$ be a function of the form (1.39). Applying the propositions from section 1.3 and formula (1.35), we conclude that $\hat{E}$ is a homogeneous distribution of degree $-(n+\mu)$ and the Fourier transform of $Q(x)\ln|x|$

is a pseudohomogeneous distribution of the same degree. The restrictions of this distributions to $\mathbb{R}^n \setminus 0$ are C^∞ functions. Thus $u(x) \in \Pi(\mu, 0)$.

Condition (1.40) follows from Theorem 1.7.1. The proof is complete. ∎

Theorem 1.8.1 shows that for $\mu < 0$ the canonical projection is the identity. For $\mu \geq 0$ the mapping (1.38) has a non-trivial kernel of dimension $\binom{\mu+n-1}{\mu}$, which is determined by (1.40). This agrees with the codimension of $\Pi^\bullet(\mu, 0)$ in $\Pi(\mu, 0)$.

By use of the notation

$$\langle u(\cdot - y), \phi(\cdot) \rangle = \langle u(\cdot), \phi(\cdot + y) \rangle \tag{1.45}$$

for translations in $\mathbb{R}^n$ we define the classes

$$\Pi^\bullet(\mu, y) := \left\{ u(\cdot - y) \Big| : u \in \Pi^\bullet(\mu, 0) \right\},$$

$$\Pi(\mu, y) := \left\{ u(\cdot - y) \Big| : u \in \Pi(\mu, 0) \right\}.$$

These are homogeneous distributions or pseudohomogeneous functions with respect to the distance vector $\vartheta = x - y$. Later, if we consider integral equations, y plays the role of the integration point and x will be the observation point. In the meantime, y is considered as a fixed point in $\mathbb{R}^n$.

1.9 Mapping Properties of Basic Operators

1.9.1 Differentiation

If y is considered as fixed, we have $\partial_j := \frac{\partial}{\partial x_j} = \frac{\partial}{\partial \vartheta_j}$. By application of the Fourier transform $\mathcal{F}_{\vartheta \to \xi}$ we obtain from (1.12) and (1.25) immediately:

$$\partial_j : \quad \Pi(\mu, y) \longmapsto \Pi(\mu - 1, y), \tag{1.46}$$

$$\partial_j : \quad \Pi^\bullet(\mu, y) \longmapsto \Pi^\bullet(\mu - 1, y). \tag{1.47}$$

1.9.2 Multiplication with Homogeneous Polynomials

If $Q(\vartheta)$ is s homogeneous polynomial in $\vartheta = x - y$ of degree ν then the multiplication with $Q(\vartheta)$, due to (1.12), acts on the Fourier image as differential operator $(-1)^\nu Q(D)$. From (1.26) follows that the distributional derivative of a canonical continuation itself is not a canonical continuation in general. Thus we have

$$Qu \in \Pi(\mu + \nu, y), \qquad \forall u \in \Pi(\mu, y). \tag{1.48}$$

For $\mu \geq -\nu$, however, the multiplication with Q does not map from $\Pi^\bullet(\mu, y)$ to $\Pi^\bullet(\mu + \nu, y)$, in general.

1.9.3 Homogeneous Elliptic Differential Operators

A homogeneous polynomial $H(\xi)$ is called *elliptic* if the condition $H(\xi) \neq 0$ is fulfilled for all $\xi \in \mathbb{R}^n \setminus 0$. Then, the order of $H(\xi)$ must be even, say, it is equal to 2κ.

For all $j \in \mathbb{N}_0$ the multiplication operator with $H(\xi)$ is a mapping from $\text{Hom}^{(j)}$ to $\text{Hom}^{(j+2\kappa)}$ which is left invertible since only zero is mapped onto zero. A left inverse can be written in the form

$$H^\sharp \left(\sum_{|\beta|=j+2\kappa} p_\beta \xi^\beta \right) = \sum_{|\alpha|=j} \sum_{|\beta|=j+2\kappa} H^\sharp_{\alpha\beta} p_\beta \, \xi^\alpha. \tag{1.49}$$

The numbers $H^\sharp_{\alpha\beta}$ depending on the two multi-indices α and β can be calculated by division of polynomials (neglecting the non-polynomial remainder in division). Foregoing permutations of the summands lead to different versions of the (non-unique) left inverse.

The image of the operator $H^\sharp$ applied to derivatives of the δ-distribution is defined by

$$\left\langle H^\sharp \delta^{(\alpha)}, \phi \right\rangle := \left\langle \delta^{(\alpha)}, H^\sharp P_{|\alpha|+2\kappa} \phi \right\rangle. \tag{1.50}$$

Since

$$\left\langle H(\xi) H^\sharp \delta^{(\alpha)}, \phi \right\rangle = \left\langle \delta^{(\alpha)}, H^\sharp P_{|\alpha|+2\kappa}(H\phi) \right\rangle = \left\langle \delta^{(\alpha)}, H^\sharp H P_{|\alpha|} \phi \right\rangle$$

the relation

$$H(\xi) H^\sharp \delta^{(\alpha)} = \delta^{(\alpha)} \tag{1.51}$$

is fulfilled for all multi-indices α. By some calculation, the explicit representation

$$H^\sharp \delta^{(\alpha)} = \sum_{|\beta|=|\alpha|+2\kappa} \frac{\alpha!}{\beta!} H^\sharp_{\alpha\beta} \delta^{(\beta)} \tag{1.52}$$

can be derived from (1.49).

The homogeneous differential operators $H(\partial)$ and $H(D)$ are assigned to $H(\xi)$ by virtue of replacing ξ_j by ∂_j or $D_j = -i\partial_j$, respectively. Due to (1.46) and (1.47), the operator $H(\partial)$ maps as follows:

$$H(\partial): \quad \Pi(\mu, y) \longmapsto \Pi(\mu - 2\kappa, y), \tag{1.53}$$

$$H(\partial): \quad \Pi^\bullet(\mu, y) \longmapsto \Pi^\bullet(\mu - 2\kappa, y). \tag{1.54}$$

1.10 The Right Inverse to Homogeneous Elliptic Differential Operators

By definition, the Fourier transform of $u \in \Pi(\mu - 2\kappa, y)$ has the form

$$\hat{u} = \mathcal{F}_{\vartheta \to \xi} u = \widehat{T}\hat{u} + \sum_{|\alpha| = \mu - 2\kappa} c_\alpha \delta^{(\alpha)}, \qquad (1.55)$$

with certain numbers c_α depending on u.

Theorem 1.10.1. *If $H(\partial)$ is a homogeneous elliptic differential operator of order 2κ then the mappings (1.53) and (1.54) are right invertible. For both mappings, the operator*

$$H^{(-1)}u := (-1)^\kappa \mathcal{F}_{\xi \to \vartheta}^{-1} \left\{ \left[\frac{1}{H(\xi)} \, \hat{u}(\xi) \Big|_{\mathbb{R}^n \setminus 0} \right]^{\bullet} + \sum_{|\alpha| = \mu - 2\kappa} c_\alpha H^{\sharp} \delta^{(\alpha)} \right\} \quad (1.56)$$

is a right inverse. The numbers c_α are determined by (1.55) and $H^{\sharp}\delta^{(\alpha)}$ is defined by (1.50) or (1.52).

Proof. It suffices to consider $y = 0$. (The general case is equivalent, up to translations). From (1.12) we obtain

$$
\begin{aligned}
\mathcal{F}\left[H(\partial)H^{(-1)}u\right] &= (-1)^\kappa \mathcal{F}\left[H(D)H^{(-1)}u\right] = (-1)^\kappa H(\xi)\mathcal{F}\left[H^{(-1)}u\right] \\
&= H(\xi)\left\{ \left[\frac{1}{H(\xi)} \, \hat{u}(\xi) \Big|_{\mathbb{R}^n \setminus 0} \right]^{\bullet} + \sum_{|\alpha| = \mu - 2\kappa} c_\alpha H^{\sharp} \delta^{(\alpha)} \right\}.
\end{aligned}
$$

Due to (1.25) and (1.51), the last expression coincides with (1.55). Thus the operator defined by (1.56) is a right inverse to (1.53).

It remains to show that the restriction of $H^{(-1)}$ to the subset $\Pi^{\bullet}(\mu - 2\kappa, 0)$ maps into $\Pi^{\bullet}(\mu, 0)$. This can be deduced from the fact that for $u \in \Pi^{\bullet}(\mu - 2\kappa, 0)$ all numbers c_α vanish and therefore the Fourier transform of $H^{(-1)}u$ consists of a canonical continuation exclusively. ∎

Denote by $H^{[-1]}(\xi)$ a pseudohomogeneous continuation of the function $H^{-1}(\xi)$ and

$$h(x) = \mathcal{F}^{-1}H^{[-1]}(\xi) \in \Pi(2\kappa - n, 0)$$

the inverse Fourier transform. Then, the pseudodifferential operator with the distribution-valued symbol $H^{[-1]}(\xi)$ has the form [35, formula(3.37)]:

$$\left(H^{[-1]}u\right)(x) = \mathcal{F}_{\xi \to x}^{-1}\left[H^{[-1]}(\xi)\hat{u}(\xi)\right] = \int_{\mathbb{R}^n} h(x - y)u(y)dy.$$

This operator, however, is apriori only defined on the space $S(\mathbb{R}^n)$, where it represents (after multiplication with $(-1)^\kappa$) a right inverse to $H(\partial)$. Formula

(1.56) gives a rule for an extension of this pseudodifferential operator to the spaces $\Pi(\mu, y)$.

We want to mention that formula (1.56) can be evaluated explicitly in many cases. Especially, for the construction of Levi functions considered in section 2.4, the following case is of importance. Let

$$T = \sum_{\max(0, -m) \leq |\beta| \leq M} q_\beta(\vartheta) D^\beta \tag{1.57}$$

be a differential operator with $m \in \mathbb{Z}$ and

$$q_\beta(\vartheta) \in Hom^{(m+|\beta|)}.$$

Due to (1.46) and (1.48), it maps from $\Pi(\mu, y)$ into $\Pi(\mu + m, y)$, $\forall \mu \in \mathbb{Z}$.

For $\nu \in \mathbb{N}$ we donote by $N_\nu(\vartheta)$ the fundamental solution of $H^\nu(\partial)$, the ν th power of the elliptic differential operator $H(\partial)$.

Lemma 1.10.1. *Let $u = TN_\nu$ for some $\nu \in \mathbb{N}$, where T is a differential operator of the form (1.57). Provided that the fundamental solutions N_ι are known for $\max(0, -m) \leq \iota - \nu - 1 \leq M$, the expression $H^{(-1)}u$ according to (1.56) can be calculated explicitly by exclusive use of algebraic manipulations and evaluations of the commutator formula (1.26).*

Proof. We desribe the algorithm. Assume that

$$\mathcal{F}_{\vartheta \to \xi} N_\nu(\vartheta) = \left[H^{-\nu}(\xi)\right]^\bullet.$$

If $\widehat{N}_\nu(\xi)$ contains additional δ-distributions, these can be handled in an obvious manner. From (1.12) and (1.25) we obtain for the Fourier transform of u:

$$\widehat{u} = \mathcal{F}(TN_\nu) = \sum_{\max(0, -m) \leq |\beta| \leq M} q_\beta(-D) \left[\frac{\xi^\beta}{H^\nu(\xi)}\right]^\bullet.$$

This expression is transformed into the form

$$\widehat{u} = \sum_{\max(0, -m) \leq |\beta| \leq M} \left[q_\beta(-D) \frac{\xi^\beta}{H^\nu(\xi)}\right]^\bullet + \cdots,$$

where $\cdots$ stands for a certain linear combination of derivatives of $\delta(\xi)$ which can be calculated by repeated application of formula (1.26). Each of the terms $q_\beta(-D)\xi^\beta H^{-\nu}(\xi)$ will be evaluated according to the quotient rule. This gives a rational function of the form $R(x)H^{-(\nu+|\beta|)}(\xi)$.

Now $\widehat{u}$ has the desired form (1.55). If $H^{(-1)}$ is applied to it, the expressions (1.52) must be evaluated to calculate $H^\sharp \delta^{(\alpha)}$. This and the inverse Fourier transform

$$\mathcal{F}^{-1} \left[\frac{R(\xi)}{H^{\nu+|\beta|+1}(\xi)}\right]^\bullet = \mathcal{F}^{-1} R(\xi) \left[\frac{1}{H^{\nu+|\beta|+1}(\xi)}\right]^\bullet = R(D) N_{\nu+|\beta|+1}(\vartheta) \tag{1.58}$$

can be performed by pure algebraic manipulations.   ∎

1.11 Local Convergent Series with Pseudohomogeneous Terms

With $\mathcal{K}_\varepsilon$ we denote the n-dimensional ball $\mathcal{K}_\varepsilon := \{x \mid : |x - y| < \varepsilon\}$.

Definition 1.11.1. *If there exists a constant $\varepsilon > 0$ such that the series*

$$f(x) = \sum_{j=\mu}^{\infty} f_j(x), \qquad f_j \in \Pi(j, y) \tag{1.59}$$

converges uniformly for all $\varepsilon' < \varepsilon$ in the n-dimensional hollow sphere $\mathcal{K}_\varepsilon \setminus \mathcal{K}_{\varepsilon'}$ then we write $f \in \Pi(\geq \mu, y)$. The largest ε fulfilling this condition is called the radius of convergence for the series (1.59).

Since for $\mu \geq 1$ the functions from $\Pi(\mu, y)$ are continuous, it is equivalently possible to require that the series

$$\sum_{j=\max(\mu,1)} f_j$$

converges uniformly in $\mathcal{K}_\varepsilon$. From this we can deduce the absolute convergence in $\mathcal{K}_\varepsilon$ by anologous arguments, like for power series.

Projections

In generalization of (1.21) we define projections P_k on $\Pi(\geq \mu, y)$ which take only the pseudohomogeneous terms of degree k:

$$(P_k u)(x) := \text{p.f.} \left[\tau^{-k} u \left(y + \tau(x - y) \right) \right], \quad \tau \to 0.$$

Clearly, the partie-finie limit must be taken only for $k \leq -n$, else it is the usual limit.

Further we define in $\Pi(\geq \mu, y)$ the projections

$$P[< k] := \sum_{j=\mu}^{k-1} P_j \tag{1.60}$$

which take all terms of degree $< k$.

1.12 Basic Operators Acting on Series with Pseudohomogeneous Terms

1.12.1 Differentiation

In a similar way, like for power series, one can show that the differentiation applied to (1.59) commutes with the summation and that the derivatives converge uniformly in all hollow spheres $\mathcal{K}_\varepsilon \setminus \mathcal{K}_{\varepsilon'}$. Then, from (1.46) follows

$$\partial_j : \quad \Pi(\geq \mu, y) \longmapsto \Pi(\geq \mu - 1, y). \tag{1.61}$$

The differentiation does not change the radius of convergence. Further, it can be verified that $f(x) \in \Pi(\geq \mu, y)$ is infinitely differentable for all points $x \in \mathcal{K}_\varepsilon \setminus \{y\}$. In the case $\mu \geq 1$, $f(x)$ is $(\mu-1)$ times continuously differentable in the point $x = y$.

1.12.2 Multiplication

If the function $F(x)$ is infinitely differentiable at the point y and the Taylor expansion converges in a ε-neighbourhood of y then the multiplication with $F(x)$ maps as follows:

$$F(x) : \Pi(\geq \mu, y) \longmapsto \Pi(\geq \mu + j, y), \quad \text{if } \left. \frac{\partial^\beta F(x)}{\partial x^\beta} \right|_{x=y} = 0, \quad \forall |\beta| \leq j. \tag{1.62}$$

The radius of convergence of the product is equal to the minimal radius of convergence of the factors.

1.12.3 Homogeneous Elliptic Differential Operators

If $H(\partial)$ is a homogeneous elliptic differential operator of order 2κ then the properties of the differentiation imply together with (1.53):

$$H(\partial) : \Pi(\geq \mu + 2\kappa, y) \longmapsto \Pi(\geq \mu, y). \tag{1.63}$$

This mapping is right invertible. Assume that the right inverse is constructed according to (1.56) and that it is applied step by step to the consecutive terms of the series. The question arises which special realization of the right inverse is to be choosen for each term in order to ensure convergence of the complete series in $\mathcal{K}_\varepsilon$.

This problem will not investigated here in detail. We only mention that such a choice is possible as can be seen from the following integral representation for a right inverse to (1.63):

$$\left(H^{(-1)}u \right)(x) = (I - P[< \mu + 2\kappa]) \int_{\mathcal{K}_\varepsilon} G(x - z)u(z)dz. \tag{1.64}$$

This representation is valid for $\mu > -n$. Hereby $G(x) \in \Pi^\bullet(2\kappa - n, 0)$ is a fundamental solution for $H(\partial)$ and ε is the radius of convergence of u.

1.12.4 Fourier Transform

Let $\chi(x) \in C_0^\infty(\mathbb{R}^n)$ be a cut-off function with following properties:

$$\chi(x) \equiv 0 \text{ for } |x| \geq 1 \quad \text{and} \quad \exists\, \theta \in (0,1) \text{ with } \chi(x) \equiv 1 \text{ for } |x| \leq \theta. \quad (1.65)$$

If $f(x) \in \Pi(\geq \mu, y)$ is a local convergent series of the form (1.59) with the radius ε of convergence, then the product

$$f_\chi(x) := \chi\left(\frac{x-y}{\varepsilon}\right) f(x)$$

is a C^∞ function in $\mathbb{R}^n \setminus y$ and at the same time a distribution or (for $\mu > -n$) a function with compact support.

Lemma 1.12.1. *For $f(x) \in \Pi(\geq \mu, y)$, the Fourier transform*

$$\widehat{f}_\chi(\xi) := \mathcal{F}_{x-y \to \xi} f_\chi(x) \qquad \left(= \int e^{-i(x-y,\xi)} f_\chi(x)dx, \quad \text{if } \mu > -n \right)$$

is an entire analytical function in $\mathbb{C}^n$. For all multi-indices $\alpha \in \mathbb{N}_0^n$ there exists a constant C_α with

$$\left| \frac{\partial^\alpha}{\partial \xi^\alpha} \widehat{f}_\chi(\xi) \right| \leq C_\alpha (1 + |\xi|)^{-(n+\mu+|\alpha|)}, \qquad \forall \xi \in \mathbb{R}^n. \quad (1.66)$$

Proof. The Fourier transform of a distribution with compact support can always be extended to an entire analytical function on $\mathbb{C}^n$ [35, Theorem 2.1]. Hence, the restriction to $\mathbb{R}^n$ is a C^∞ function and it suffices to prove the inequality (1.66) for $|\xi| \geq 1$. We consider at first the case $\alpha = 0$.

Pseudohomogeneous functions of degree μ are locally square integrable for $\mu > -n/2$. Hence, for all multi-indices β with $|\beta| < \mu + n/2$ holds: $\partial^\beta f_\chi(x) \in L^2$. Together with (1.12) and the mapping property $\mathcal{F}: L^2 \longrightarrow L^2$ this leads to $\xi^\beta \widehat{f}_\chi(\xi) \in L^2$. This means that $\widehat{f}_\chi(\xi)$ decays faster than $|\xi|^{-(|\beta|+n/2)}$ for $|\xi| \to \infty$. Since $\widehat{f}_\chi$ is continuous, we obtain an estimate of the form

$$\left| \widehat{f}_\chi(\xi) \right| \leq C|\xi|^{-\lambda}, \qquad \forall \mu > -\frac{n}{2}, \quad \forall |\xi| \geq 1. \quad (1.67)$$

Hereby $\lambda = \mu + n$ or $\lambda = \mu + n - 1/2$ for even or odd n, respectively.

If μ is an abitrary integer, we choose an integer $N \geq \mu$, $N > -n/2$ and decompose f_χ into three summands:

$$f_\chi(x) = \sum_{j=\mu}^N f_j(x) + \left[\chi\left(\frac{x-y}{\varepsilon}\right) - 1 \right] \sum_{j=\mu}^N f_j(x) + \chi\left(\frac{x-y}{\varepsilon}\right) \sum_{j=N+1}^\infty f_j(x).$$
$$(1.68)$$

The Fourier transform of each function $f_j \in \Pi(j, y)$ is the continuation of a homogeneous C^∞ function of degree $-(n+j)$ and fulfills (due to $j \geq \mu$) the inequality (1.66) for $|\xi| \geq 1$. The second summand in (1.68) is a C^∞

function of polynomial growth. It is well known that its Fourier transform decays (together with all derivatives) for $\xi \to \infty$ faster than $|\xi|^{-k}$, $\forall k \in \mathbb{N}$. The Fourier transform of the third summand in (1.68) can be estimated by (1.67).

Therefore, (1.66) is proved for $\alpha = 0$. The assertion for $\alpha \neq 0$ can be reduced to this case by use of the mapping property (1.61).   ∎

1.13 The Parity Condition

For simplicity we use the same notation $u = u(x)$ for distributions and functions. Especially, the distribution $u(-x)$ is defined by

$$\langle u(-x), \phi(x) \rangle = \langle u(x), \phi(-x) \rangle .$$

Definition 1.13.1. *Let $u \in S'(\mathbb{R}^n)$ be a pseudohomogeneous distribution of degree μ. We say that u fulfills the "parity condition with respect to zero" if*

$$u(-x) = (-1)^{n+\mu} u(x). \tag{1.69}$$

We say that a distribution $u \in \Pi(\mu, y)$ fulfills the "parity condition" if u is obtained by the translation (1.45) from a distribution fulfilling the parity condition with respect to zero. If u is a function, this means:

$$u(y - x) = (-1)^{n+\mu} u(x - y).$$

With $\Psi^\bullet(\mu, y)$ or $\Psi(\mu, y)$ we denote the subset of all $u \in \Pi^\bullet(\mu, y)$ or $u \in \Pi(\mu, y)$, respectively, fulfilling the parity condition. Accordingly, $\Psi(\geq \mu, y)$ denotes the subset of all series of the form (1.59) where all elements fulfill the parity condition.

That $u \in \Pi(\mu, 0)$ fulfills the parity condition can be formulated equivalently with the help of the Fourier transform:

$$\widehat{u}(-\xi) = (-1)^k \widehat{u}(\xi). \tag{1.70}$$

Here, $k = -(n + \mu)$ denotes the degree of homogeneity of $\widehat{u}(\xi)$. As can be seen, the Fourier transform itself fulfills the parity condition if and only if n is even.

Lemma 1.13.1. *If the space dimension n is odd then:*

(i) $\Psi(\mu, y) = \Psi^\bullet(\mu, y) \subset \Pi^\bullet(\mu, y)$.

(ii) All functions (resp. distributions) from $\Psi(\mu, y)$ are homogeneous in ϑ. The logarithmic terms vanish.

Proof. (i) Let $\mu \geq 0$ and $u \in \Psi(\mu, y)$. We have to shown that this implies $u \in \Pi^{\bullet}(\mu, y)$. The Fourier transform of u has the representation (1.55) with $\kappa = 0$. This relation is re-arranged as follows:

$$\widehat{v}(\xi) := \widehat{u}(\xi) - \left[\widehat{u}(\xi)\Big|_{\mathbb{R}^n \setminus 0}\right]^{\bullet} = \sum_{|\alpha|=\mu} c_\alpha \delta^{(\alpha)}. \tag{1.71}$$

Due to (1.70), there holds $\widehat{u}(-\xi) = (-1)^{n+\mu}\widehat{u}(\xi)$. The canonical continuation fulfills the same relation, which can be seen immediately from (1.17). A consideration of the left-hand side of (1.71) leads to $\widehat{v}(-\xi) = (-1)^{n+\mu}\widehat{v}(\xi)$ while a consideration of the right-hand side implies $\widehat{v}(-\xi) = (-1)^{\mu}\widehat{v}(\xi)$. For odd numbers n this is the opposite sign. Hence, $\widehat{v} = 0$ and $u \in \Pi^{\bullet}(\mu, y)$, q. e. d.

(ii) For $\mu \geq 0$ and $u \in \Psi(\mu, y)$ the distribution

$$w(\xi) := \left(\frac{\xi}{|\xi|}\right)^{\beta} \widehat{u}(\xi)$$

fulfills for $|\beta| = \mu$ the relation $w(-\xi) = (-1)^{\mu}w(\xi)$. Thus, the conditions (1.24) are fulfilled for odd numbers n, and only the homogeneous parts are left over for $\widehat{u}$ and u.   ∎

Lemma 1.13.2.

(i)   The mapping properties (1.46), (1.47), (1.48), (1.53), (1.54), (1.61), (1.62), (1.63) of the basic operators are preserved if the letter Π is replaced by Ψ.

(ii) The operator (1.56) preserves the parity condition.

(iii) The fundamental solution of a homogeneous elliptic differential operator of order 2κ belongs to the class $\Psi^{\bullet}(2\kappa - n, y)$, i. e., it fulfills always the parity condition.

Proof. (i) Each homogeneous polynomial $Q(x)$ of degree ν satisfies the relation $Q(-x) = (-1)^{\nu}Q(x)$. This gives immediately the assertions for the operators of multiplication with homogeneous polynomials. By use of the Fourier transform and (1.12) they can be carried over to the differential operators.

(ii) If $H(\xi)$ is an elliptic polynomial then $H(-\xi) = H(\xi)$. The assertion for the operator $H^{(-1)}$ can immediately be deduced from (1.56) if we take into account that for odd numbers n all coefficients c_α vanish by virtue of Lemma 1.13.1.

(iii) The δ-distribution belongs to $\Pi^{\bullet}(-n, 0)$ and fulfills the parity condition since $\delta(-x) = \delta(x)$. The fundamental solution is generated by application of the operator (1.56) which preserves the parity condition, as we have already shown.   ∎

Now we are able to reproduce the classical results concerning the structure of fundamental solutions for homogeneous elliptic differential operators as they can be found e. g. in [72] and [73]. The relation $\mathcal{F}(\delta) = 1$ implies $\delta \in \Pi^\bullet(-n, 0)$ and $\delta(x - y) \in \Pi^\bullet(-n, y)$. Thus

$$\mathcal{F}^{-1}_{\xi \to \vartheta} \left[\frac{1}{H(\xi)} \right]^\bullet \in \Pi^\bullet(2\kappa - n, y) \tag{1.72}$$

is a fundamental solution for $H(\partial)$. More explicit representations for fundamental solutions in the case $n = 2$ can be found in [45] and [93].

2. Levi Functions for Elliptic Systems of Partial Differential Equations

All shell models devoted to the static, linear elastic case, especially the shell equations considered in part II, belong to the class of elliptic systems in the sense of Douglis and Nirenberg, they are mixed-order systems as investigated e. g. in [29]. Chapter 2 collects some basic notations and results for such systems. Further, a general algorithm is developed for the construction of Levi functions.

Sections 2.1–2.3 collect classical definitions and assertions, only the concept of the degree of a Levi function is newly introduced here.

2.1 Definitions of Ellipticity in the Sense of Douglis and Nirenberg

We consider a linear system of partial differential equations

$$\sum_{k=1}^{m} A^{jk}(x, \partial) u_k(x) = -p_j(x), \qquad j = 1, \ldots, m, \tag{2.1}$$

in a domain $\Omega \subset \mathbb{R}^n$. Suppose that there exist two m-tuples $s_1, \ldots, s_m$ and $t_1, \ldots, t_m$ of integers such that all operators

$$A^{jk}(x, \partial) = \sum_{|\beta|=0}^{s_j+t_k} a_\beta^{jk}(x) \partial^\beta \tag{2.2}$$

are polynomials of degree

$$\deg A^{jk}(x, \partial) \le s_j + t_k \tag{2.3}$$

in $\partial = (\partial_1, \ldots, \partial_n)$. The coefficients $a_\beta^{jk}(x)$ are assumed to be real-valued. (All investigations can be carried over to the complex-valued case without problems.) Requirements concerning the smoothness of these coefficients are specified later. For the following definitions it suffices to assume continuity.

In the sequel we need the notations

$$s_0 := \max\{s_1, \ldots, s_m\}, \quad t_0 := \max\{t_1, \ldots, t_m\}, \quad s := \sum s_j, \quad t := \sum t_k.$$

The *principal part* $A_0(x, \partial)$ of the operator $A(x, \partial)$ is a $m \times m$ matrix with the entries

$$A_0^{jk} := \sum_{|\beta| = s_j + t_k} a_\beta^{jk}(x) \partial^\beta. \tag{2.4}$$

If the order of A^{jk} is less than $s_j + t_k$, we set $A_0^{jk} = 0$.

The *characteristic matrix* $A_0(x, \xi)$ is obtained if the differentiation ∂ is replaced formally by the n-tuple $\xi = (\xi_1, \ldots, \xi_n)$ of complex numbers. The determinant

$$H(x, \xi) := \det[A_0(x, \xi)] \tag{2.5}$$

is a homogeneous polynomial in ξ of degree $s + t$. Suppose that $H(x, \xi)$ does not vanish, the number $s + t$ is called the *order* of the system (2.1). If $H(x, \xi) \equiv 0$, the numbers s_j and t_k must be changed appropriately.

Definition 2.1.1. *If the numbers s_j and t_k can be choosen in such a way that the condition*

$$H(x, \xi) \neq 0, \qquad \forall x \in \Omega, \ \forall \xi \in \mathbb{R}^n \setminus 0 \tag{2.6}$$

is fulfilled then the operator $A(x, \partial)$ is called "elliptic in the sense of Douglis and Nirenberg".

The ellipticity implies that the order is even: $s + t =: 2\kappa$. In the case $\kappa = 0$, the system (2.1) can be solved explicitly and no degrees of freedom remain for boundary conditions [29]. Thus we assume $\kappa \geq 1$ in what follows.

In connection with the Fourier transform it seems to be more natural to write the operator (2.1) as a polynomial in $D_j = -i\partial_j$ and to substitute all D_j by real numbers ξ_j when the characteristic matrix is formed. The definition of ellipticity is not influenced by this modification, for the following definition of strong ellipticity, however, this is of importance.

Definition 2.1.2. *An elliptic operator $A(x, \partial)$ is called "strongly elliptic" if all sums $s_j + t_j$ are even and if there exist two constants $c > 0$, $\phi \in [0, 2\pi)$ independent of x such that the inequality*

$$\mathfrak{Re}\left[e^{i\phi} \sum_{j,k=1}^m A_0^{jk}(x, i\xi)\eta_j \overline{\eta_k} \right] \geq c \sum_{j=1}^m |\xi|^{s_j + t_j} |\eta_j|^2, \qquad \forall \eta \in \mathbb{C}^n, \ \forall \xi \in \mathbb{R}^n \tag{2.7}$$

is fulfilled.

A more general concept of strong ellipticity was developed in [122]. It says that a pseudodifferential operator is strongly elliptic if it is the sum of a definite and a compact operator. Due to this concept, a regular matrix-valued function enters enters into (2.7) instead of the constant factor $e^{i\phi}$.

2.2 Fundamental Solutions and Levi Functions

A $m \times m$ matrix $L(x, y)$ of functions depending besides on x on n parameters $y = (y_1, y_2, \ldots, y_n)^\top$ is called a *Levi function*[1], if the equation

$$A(x, \vartheta)L(x, y) = \delta(x - y) \cdot I_{m \times m} - R(x, y) \qquad (2.8)$$

is fulfilled in the distributional sense. Here, $I_{m \times m}$ denotes the unit matrix and δ is Dirac's δ-distribution in $\mathbb{R}^n$. Further one requirs that for all points $x \neq y$ the functions[2] in the kth row of $L(x, y)$ are at least $s_0 + t_k$ times continuously differentiable with respect to x and y. This agrees with the maximal order of differentiation applied to them.

Concerning the remainder $R(x, y)$ we assume that it has a higher degree of homogeneity than the δ-distribution $\delta(x - y)$ in the vicinity of the hyperplane $x = y$. This means, there exist three constants $C > 0$, $\varepsilon > 0$ and $\lambda > 0$ such that for each row index $j = 1 \ldots m$ and each column index $k = 1 \ldots m$ the corresponding entry of the matrix $R(x, y)$ is bounded by

$$|R_{jk}(x, y)| \leq C|x - y|^{\lambda - n}, \qquad \text{for } 0 < |x - y| < \varepsilon. \qquad (2.9)$$

Definition 2.2.1. *If the inequality (2.9) can be replaced by the stronger condition*

$$|R_{jk}(x, y)| \leq |U(x - y)|, \qquad \text{for } 0 < |x - y| < \varepsilon$$

with a certain pseudohomogeneous function $U(\vartheta)$ of degree $\mu - n$ ($\mu \in \mathbb{N}$) then we call $L(x, y)$ a "Levi function of degree μ". As usual, a Levi function with a vanishing remainder $R(x, y)$ is called a "fundamental solution".

2.3 Existence Results for Fundamental Solutions

If the existence is ensured for all pairs of points x, y from the considered domain Ω then we speak of *global existence*. *Local existence* means only that for a certain constant $\delta > 0$, independent of x and y, the fundamental solution exists for all pairs of points $x, y \in \Omega$ with $|x - y| < \delta$.

For elliptic systems in the sense of Douglis und Nirenberg the following results are known:

1. A Levi function exists globally in each case (see section 2.4).

---

[1] The name "parametrix" is often used for such functions as well. It was introduced by D. Hilbert [59, 60] for functions fulfilling (2.8) and, additionally, the condition $L(x, y) = L(y, x)$. Very often, however, the name "paramerix" is used in the literature for a Levi function with a C^∞ remainder $R(x, y)$ or for a regularizing operator. With the name "Levi function" we follow the terminology used e. g. in [88] and [89].

[2] In general, these are distributions. For $x \neq y$, however, they have representations as functions. That is why we call them functions, for simplicity.

2. In the special case of constant coefficients, the global existence of a fundamental solution is ensured [29].

3. For scalar operators of the second order $A(x, \partial) = \sum_{|\beta|=0}^{2} a_\beta(x) \partial^\beta$ with $|\beta|$ times Hölder-continuously differentiable coefficients $a_\beta(x)$, a fundamental solution exists globally. This assertion is formulated as Theorem 19, VIII in [89], however, the proof given there is incomplete. A complete proof under slightly stronger assumptions concerning the smoothness of the coefficients can be found in [82]. Formerly, G. Fichera [37] proved such a result for the strong elliptic case.

4. In general, the system (2.1) with Hölder-continuous coefficients possesses locally a fundamental solution, the global existence can be ensured only under additional conditions for the operator and the domain [89, sect. 55.B and 55.C].

5. Already in the scalar case $m = 1$, examples of elliptic operators with C^∞ coefficients are known for which no fundamental solution exists globally [89, sect. 52.B].

It is probably possible to show global existency of a fundamental solution for the shell equations considered in part II, if the additional information from Theorem 4.11.1 is used, namely that the operator is positive definite in appropriate Hilbert spaces. This is a condition which regards the complete operator and not only the principal part. Since the existence of a fundamental solution plays no role in what follows, we won't investigate this question.

2.4 A General Algorithm for the Construction of Levi Functions

2.4.1 Assumptions

Suppose $A(x, \partial)$ to be elliptic of order $2\kappa \geq 2$. For simplicity we assume that all coefficients $a_\beta^{jk}(x)$ are real-analytic in the closure $\overline{\Omega}$. This means, there exists a domain $\Omega_1 \supset \Omega$ where the coefficients possess an analytic continuation. This assumption can be weakened and the algorithm described below can be applied in cases of lower smoothness as well. Levi functions of arbitrary degree, however, can not be achieved then. Instead, upper bounds for the reachable degrees exist (see section 2.6).

2.4.2 The Steps of the Algorithm

1st step. The operator $A = A(x, \partial)$ is split into the difference

$$A = A_0 - T, \tag{2.10}$$

where $A_0 := A_0(y, \partial)$ denotes the principal part with frozen coefficients and T denotes the complete remainder $T := A_0 - A$.

2nd step. Since $A(x, \partial)$ is assumed to be elliptic, the operator

$$H(y, \partial) = \det [A_0(y, \partial)]$$

is a homogeneous elliptic differential operator of degree 2κ with constant coefficients, by definition. As described in section 1.12, a right inverse $H^{(-1)}$ can be constructed which maps from $\Psi(\geq \mu, y)$ into $\Psi(\geq 2\kappa + \mu, y)$.

3rd step. With $B(x, \xi)$ we denote the cofactor matrix of $A_0(x, \xi)$. This means, $B_{jk}(x, \xi)$ is equal to $(-1)^{j+k}$ multiplied with the determinant of $A_0(x, \xi)$ after deleting the kth row and the jth column. According to Cramer's rule there holds

$$A_0(x, \xi) \cdot B(x, \xi) = B(x, \xi) \cdot A_0(x, \xi) = H(x, \xi) \cdot I_{m \times m}. \tag{2.11}$$

Thus

$$A_0^{(-1)} := B(y, \partial) \, H^{(-1)}(y, \partial) \tag{2.12}$$

is a right inverse to A_0. A right inverse to $A = A(x, \partial)$ can then be written in the form

$$A^{(-1)} = \left(I - A_0^{(-1)} T\right)^{-1} A_0^{(-1)}, \tag{2.13}$$

provided that the operator $I - A_0^{(-1)} T$ is invertible. The question whether or not this is really true, is not of importance here. Relation (2.13) has only formal character.

4th step. The (fictitious) inverse of $I - A_0^{(-1)} T$ in (2.13) is replaced by the Neumann series. This leads to the iteration scheme

$$N^{(0)}(x, y) = A_0^{(-1)} [\delta(x - y) I_{m \times m}] \tag{2.14}$$

$$N^{(\iota+1)}(x, y) = A_0^{(-1)} T N^{(\iota)}(x, y), \qquad \iota = 0, 1, 2, \ldots \tag{2.15}$$

which generates a sequence of matrices $N^{(0)}, N^{(1)}, \ldots$. The sums

$$L^{(\mu)}(x, y) := \sum_{\iota=0}^{\mu} N^{(\iota)}(x, y) = A_0^{(-1)} \sum_{\iota=0}^{\mu} \left(T A_0^{(-1)}\right)^{\iota} [\delta(x - y) \cdot I_{m \times m}] \tag{2.16}$$

form Levi functions for the operator $A(x, \partial)$ for sufficiently large numbers μ. This will be shown in the next section.

2.4.3 Practical Realization

The operators $B(y, \partial)$ and T contain only multiplications and differentiations. The crucial step is the application of the operator $H^{(-1)}(y, \partial)$ which requires an evaluation of formula (1.56). Let us have a closer look at this problem.

The numbers c_α can be calculated explicitly according to Corollary 1.6.1 and Theorem 1.7.1. If u is the δ-distribution (which is especially the case in the first step (2.14)), the inverse Fourier transform of $[1/H(\xi)]^\bullet$ is nothing else but the fundamental solution of a homogeneous elliptic differential operator.

The entries of the operator matrix $TB(y,\partial)$ can be expanded into a series of differential opertors of the form (1.57) if the coefficient $a_\beta^{jk}(x)$ are replaced by the corresponding Taylor expansions. As decribed in Lemma 1.10.1, the application of $A_0^{(-1)}$ to each matrix entry can be evaluated explicitly if the fundamental solutions of the powers of $H(y,\partial)$ are known. Moreover, the obtained terms can be represented as the application of a differential operator to a fundamental solution. This situation remains unchanged if the operators $TB(y,\partial)$ and $A^{(-1)}$ are applied repeatedly.

The problem reduces to the calculation of fundamental solutions for homogeneous elliptic differential operators with constant coefficients. In each case, the method of *plane wave decomposition* can be applied and the fundamental solution can be represented by a formula which contains a $(n-2)$-dimensional integral [73, 46]. A closed formula without integrations can always be obtained for $n \leq 2$. For $n = 1$, the inverse $H^{(-1)}$ can be written in explicit form (see section 2.8). In the case $n = 2$ it is possible to use integrations in the complex plane for an explicit evaluation of $H^{(-1)}$ (see section 5.5). For $n \geq 3$, closed formulas can be found if $H(y,\partial)$ can be transformed by a substitution of variables to the Laplacian or powers of it.

The calculation can be simplified in case $n = 2$ if first a symbolic representation of $H^{(-1)}u$ is derived from (1.56) and thereafter the undetermined coefficients are calculated by the method of comparison of coefficients. Symbolic manipulation with computer algebra appears here as a very helpful tool for performing all the calculations.

2.5 Local Smoothness Properties

We consider the first term

$$N^{(0)} = B(y,\partial)H^{(-1)}(y,\partial)\,\delta(x-y)I_{m\times m}.$$

Due to (1.72) and Lemma 1.13.2 there holds

$$H^{(-1)}(y,\partial)\delta(x-y) \in \Psi^\bullet(2\kappa - n, y). \tag{2.17}$$

The particular entries of the matrix $B(y,\partial)$ are differential operators of order

$$\text{ord } B_{jk}(y,\partial) \leq 2\kappa - s_k - t_j \tag{2.18}$$

as can be seen from the construction of the cofactor matrix. If this operator is applied to the diagonal matrix with the entries (2.17), we obtain due to Lemma 1.13.2:

$$L_{jk}^{(0)} = N_{jk}^{(0)} \in \Psi(\geq s_k + t_j - n, y). \tag{2.19}$$

For the remainder

$$R^{(\mu)}(x,y) := \delta(x-y)I_{m\times m} - A(x,\partial)L^{(\mu)}(x,y), \qquad (2.20)$$

we derive from (2.16) the relations

$$A(x,\partial)L^{(\mu)}(x,y) \;=\; \delta(x-y)\cdot I_{m\times m} - A_0 N^{(\mu+1)}(x,y),$$
$$R^{(\mu)} \;=\; A_0 N^{(\mu+1)}. \qquad (2.21)$$

The operator $T = A_0 - A$ contains at the intersection of the jth row with the kth column a differential operator of degree $s_j + t_k$, the coefficients of which vanish for $x = y$ and, additionally, differential operators of lower order. (Of course it is possible that only operators of lower order occur.) The mapping properties from Lemma 1.13.2 imply:

$$T_{jk} : \Psi(\geq l, y) \longmapsto \Psi(\geq l+1-s_j-t_k, y). \qquad (2.22)$$

This gives for the particular entries of the matrix $R^{(0)} = A_0 N^{(1)} = TL^{(0)}$ the relation

$$R^{(0)}_{ji} = \sum_{k=1}^{m} T_{jk}L^{(0)}_{ki} \in \Psi(\geq 1 + s_i - s_j - n, y). \qquad (2.23)$$

This formula shows that in general $L^{(0)}$ is not a Levi function for the operator $A(x,\partial)$. The assumptions of differentiability are fulfilled since in each sufficiently small neighbourhood of y, excluding the point y itself, the function $L^{(0)}$ is infinitely differentiable. However, if the numbers s_j are significantly different from each other, condition (2.9) may be violated.

Nevertheless, the minimal degree of pseudohomogeneity on each position of the matrices $R^{(\mu)}$ is raised in each iteration step (2.15) at least by one. The operator

$$\left(A_0^{(-1)}T\right)_{ij} = \sum_{k=1}^{m} B_{ik}(y,\partial)H^{(-1)}(y,\partial)T_{kj}$$

in the ith row and jth column of $A_0^{(-1)}T$ has the mapping property

$$\left(A_0^{(-1)}T\right)_{ij} : \quad \Psi(\geq l, y) \longmapsto \Psi(\geq l+1+t_i-t_j, y), \qquad (2.24)$$

as can be verified from (2.18) and (2.22) . Induction with respect to ι, starting with (2.19), leads to

$$N^{(\iota)}_{jk} \in \Psi(\geq \iota + s_k + t_j - n, y). \qquad (2.25)$$

Together with (2.21) this implies

$$R^{(\mu)}_{jk} \in \Psi(\geq \mu + 1 + s_k - s_j - n, y). \qquad (2.26)$$

The last assertion can be formulated as follows.

Theorem 2.5.1. *For $\mu = \lambda - 1 + \max_{j,k}(s_j - s_k)$ and $\lambda \geq 1$, the function $L^{(\mu)}$ is always a Levi function of degree $\geq \lambda$.*

2.6 Cutting the Higher Order Terms by Projections

If one is apriori interested in a Levi function of a pre-determined degree λ, the computational expense can be reduced significantly if in each step only those pseudohomogeneous terms which are really needed are taken into consideration. For this purpose we introduce the diagonal matrices $\mathbb{P}_q$ defined for all $q \in \mathbb{Z}$ as follows: at the k-th position ($k = 1, \ldots, m$) stands the projection $P[< q + s_0 - s_k - n]$ as it is defined by (1.60).

Theorem 2.6.1. *For $\mu = \lambda - 1 + \max_{j,k} (s_j - s_k)$ and $\lambda \geq 1$ the matrix $N^{(\mu)}$ keeps a Levi function of degree $\geq \lambda$ also if the iteration step (2.15) is replaced by*

$$N^{(\iota+1)}(x, y) = A_0^{(-1)} \mathbb{P}_\lambda T N^{(\iota)}(x, y), \qquad \iota = 0, 1, 2, \ldots \qquad (2.27)$$

Doing so, the iteration sceme terminates at the latest after $\mu + 1$ steps.

Proof. Analogously to (2.24) the mapping property

$$\left(T A_0^{(-1)} \right)_{ij} : \quad \Psi(\geq l, y) \longmapsto \Psi(\geq l + 1 + s_j - s_i, y)$$

can be proved, which implies

$$\mathbb{P}_q T A_0^{(-1)} = \mathbb{P}_q T A_0^{(-1)} \mathbb{P}_{q-1}. \qquad (2.28)$$

This formula represents the fact that the operator $T A_0^{(-1)}$ smoothes everything by one degree.

For the matrices generated by the iteration scheme (2.27) consequently holds (whereby we write δ instead of $\delta(x - y) I_{m \times m}$ for simplicity):

$$N^{(\iota)} = A_0^{(-1)} \left(\mathbb{P}_\lambda T A_0^{(-1)} \right)^\iota \delta = A_0^{(-1)} \mathbb{P}_\lambda \left(T A_0^{(-1)} \right)^\iota \delta, \qquad (2.29)$$

$$L^{(\mu)} = \sum_{\iota=0}^{\mu} N^{(\iota)} = A_0^{(-1)} \mathbb{P}_\lambda \sum_{\iota=0}^{\mu} \left(T A_0^{(-1)} \right)^\iota \delta,$$

$$R^{(\mu)} = \delta - (A_0 - T) L^{(\mu)} = \delta - \left(I - T A_0^{(-1)} \right) \mathbb{P}_\lambda \sum_{\iota=0}^{\mu} \left(T A_0^{(-1)} \right)^\iota \delta.$$

It can be seen that the term $\mathbb{P}_\lambda R^{(\mu)}$ contains exactly the same as if it was generated by the sceme (2.15). In this case relation (2.26) holds, which implies that the identity $\mathbb{P}_\lambda R^{(\mu)} = 0$ is fulfilled for $\mu \geq \lambda - 1 + \max_{j,k} (s_j - s_k)$. This means that $R^{(\mu)}$ contains only pseudohomoheneous terms of degree $\geq \lambda + s_0 - s_j - n$. Since $s_0 - s_j \geq 0$, we obtain the result that $L^{(\mu)}$ is a Levi function of degree $\geq \lambda$.

Due to (2.28) we can write (2.29) likewise in the form

$$N^{(\nu)} = A_0^{(-1)} \left[\prod_{j=0}^{\nu-1} \left(\mathbb{P}_{\lambda-j} T A_0^{(-1)} \mathbb{P}_{\lambda-j-1} \right) \right] \delta.$$

For $\nu > \lambda + \max_{j,k} (s_j - t_k)$ we have $\mathbb{P}_{\lambda-\nu}\delta = 0$ and hence $N^{(\nu)} = 0$. This means that the procedure terminates. ∎

The insertion of the projections into the iteration sceme has the effect that all coefficients of the operators $A(x, \partial)$ and T are in fact replaced by cut Taylor expansions about the point y. In which order this cut is done for each particular function, can easily be determined by apriori. We see that the described construction for Levi functions of a prescribed degree works as well if the assumption of real-analytic coefficients is weakened to the assumption that they are as often continuously differentiable as desired.

Furthermore, we remark that it could be also of interest for the applications (i. e. for the transformation of boundary value problems into integral equations) to construct *Levi functions of non-uniform degree*. This means that the exponent μ occuring in (2.9) can be different for different indices j, k. Even in this case the corresponding projections can be applied and the iteration terminates as well after a finite number of steps.

2.7 Pseudohomogeous Series Expansions of Fundamental Solutions

Remark 2.7.1. If the series

$$G(x, y) = \sum_{\iota=0}^{\infty} N^{(\iota)}(x, y) \tag{2.30}$$

converges, then the limit represents a fundamental solution of the operator A.

It is known that the considered operator $A(x, \partial)$ possesses always locally a fundamental solution (see section 2.3). Further it can be shown that the application of the projections $P[< \mu]$ to this local fundamental solution coincides with $L^{(\mu-1)}$ modulo smooth solutions of the homogeneous equation $Au = 0$. This allows the conclusion that the right inverse $H^{(-1)}$ can always be determined in such a way that the series (2.30) converges in a sufficiently small neighbourhood of the point y.

The question of the size of this neighbourhood is closely related to the question of the radius of convergence for local power series solutions of Cauchy problems according to Cauchy-Kowalewskaja's theorem. The methods known from there can be carried over to the question considered here. This analogy was already noted and used by J. Hadamard [55].

Since the aim of this book is the construction of Levi functions and not the construction of fundamental solutions, this problem will not be investigated in general. Only a special, one-dimensional example will be considered in the next section.

2.8 The Example of an Ordinary Differential Operator

The general construction principle for Levi functions is now demonstrated for the case $m = n = \kappa = 1$. We consider the ordinary differential operator of the second order

$$A(x, \partial) = a_2(x)\frac{d^2}{dx^2} + a_1(x)\frac{d}{dx} + a_0(x) \tag{2.31}$$

in a closed interval $\overline{\Omega} \subset \mathbb{R}^1$. All coefficients are assumed to be real-valued and continuous in $\overline{\Omega}$. Futhermore, a_2 is assumed to be Lipschitzcontinuous in $\overline{\Omega}$. The ellipticity condition for the operator (2.31) takes the form: $a_2(x) \neq 0$, $\forall x \in \Omega$. Further we require (without loss of generality) that a_2 does not vanish at the endpoints of the interval.

According to (2.10), the operator (2.31) is split into the difference $A_0 - T$ with

$$A_0 = a_2(y)\frac{d^2}{dx^2}, \qquad T = [a_2(y) - a_2(x)]\frac{d^2}{dx^2} - a_1(x)\frac{d}{dx} - a_0(x).$$

The right inverse (2.12) has now the form

$$(A_0^{(-1)}f)(x) = \frac{1}{a_2(y)}\int_y^x (x - t)f(t)dt.$$

One obtains

$$L^{(0)}(x, y) = N^{(0)}(x, y) = \frac{|x - y|}{2a_2(y)}$$

as Levi function of first degree. Levi functions of higher degree can be calculated explicitly by use of the iteration scheme (2.15) or (2.27) and subsequent summation. The next term has the form

$$N^{(1)}(x, y) = -\frac{\operatorname{sgn}(x - y)}{2\left[a_2(y)\right]^2}\int_y^x (x - t)\left[a_1(t) + (t - y)a_0(t)\right]dt.$$

Under the additional assumption that a_1 is Lipschitz continuous, we obtain

$$P[< 3]L^{(1)}(x, y) = \frac{|x - y|}{2a_2(y)}\left[1 - \frac{a_1(y)}{2a_2(y)}(x - y)\right] \tag{2.32}$$

as Levi function of second degree.

To investigate the convergence of the series (2.30), we use an idea for the proof of Cauchy-Kowalewskaja's theorem as it can be found in [131]. This method uses Banach's contraction principle and integral operators of Volterra type and it can be applied as well in more general cases.

Theorem 2.8.1. *In the case $m = n = \kappa = 1$, the series (2.30) converges for all x from the maximal open subinterval which contains the point y and in which the inequality*

$$\frac{a_2(x)}{a_2(y)} < 2 \tag{2.33}$$

is fulfilled.

Proof. As can be seen from (2.16), the series (2.30) converges if the operator $T A_0^{(-1)}$ is a contraction. Now, this operator has the form

$$
\left(T A_0^{(-1)} f \right)(x)
$$
$$
= \frac{a_2(y) - a_2(x)}{a_2(y)} f(x) - \frac{a_1(x)}{a_2(y)} \int_y^x f(t)\,dt - \frac{a_0(x)}{a_2(y)} \int_y^x (x - t) f(t)\,dt.
$$

It is the sum of a multiplication operator and an integral operator V of Volterra type. Due to the assumption, the absolute value of the kernel of V is bounded by:

$$C := \max_{x,y,t \in \overline{\Omega}} \left| \frac{a_1(x) + (x - t) a_0(x)}{a_2(y)} \right|. \tag{2.34}$$

For all $\varepsilon > 0$ and all $y \in \mathbb{R}^1$ let

$$E(\varepsilon, y) := \left\{ f(x) \in C(\overline{\Omega}) \;\middle|\; \|f\|_{(\varepsilon,y)} < \infty \right\}$$

denote the weighted Banach space of all continuous functions on $\overline{\Omega}$ with the norm

$$\|f\|_{(\varepsilon,y)} := \max_{x \in \overline{\Omega}} \frac{|f(x)|}{\cosh\left(\frac{x-y}{\varepsilon}\right)}.$$

This norm is equivalent to the usual maximum norm in $C(\overline{\Omega})$ if the interval Ω is bounded. From the relations

$$\frac{1}{\cosh\left(\frac{x-y}{\varepsilon}\right)} \left| \int_y^x \cosh\left(\frac{t-y}{\varepsilon}\right) dt \right| = \varepsilon \left| \tanh\left(\frac{x-y}{\varepsilon}\right) \right| < \varepsilon, \quad \forall x \in \overline{\Omega},$$

$$\|(Vf)(x)\|_{(\varepsilon,y)} \le C \max_{x \in \overline{\Omega}} \frac{1}{\cosh\left(\frac{x-y}{\varepsilon}\right)} \int_y^x |f(t)|\,dt \le C\varepsilon \|f\|_{(\varepsilon,y)}$$

follows that the norm of the opertor V can be made as small as desired by an appropriate choice of ε. The convergence or divergence behaviour of the series (2.30) depends only on the operator of the multiplication with the function

$$\tilde{a}_2(x,y) := \frac{a_2(y) - a_2(x)}{a_2(y)}. \tag{2.35}$$

If the absolute value of $\tilde{a}_2(x,y)$ is less or greater than 1, the Neumann series converges or diverges, respectively. This behaviour is not changed by a sufficiently small pertubation.

It can easily be verified that the series (2.30) has an alternating behaviour for $\tilde{a}_2(x,y) = -1$. Values ≥ 1 cannot be taken by the function (2.35), due to the ellipticity condition. Hence, the series (2.30) converges for all x from the open interval on which the function (2.35) is greater than -1 (excluding the endpoints of the interval). The theorem is proved.   ∎

To investigate the convergence of the series (2.30) in cases where the inequality (2.33) if fulfilled for x and y, however not for all intermediate points, requires the distinction of several special cases. This will not be investigated here.

Remark 2.8.1. Of course it is possible to divide the original operator (2.31) by $a_2(x)$ in order to achieve an overall convergent series representation for the fundamental solution. This method, however, cannot be carried over to the case $n > 1$.

2.9 Global Smoothness Properties

Within this section, the domain Ω is assumed to be bounded.

Definition 2.9.1. *By $\Psi\overline{\Omega}(\nu)$ we denote the class of all functions $g(x,y)$ (or distributions for $\nu \leq -n$) with the following properties:*

1. *$g(x,y)$ is a C^∞ function on $\{(x,y) \in \Omega \times \Omega | \, x \neq y\}$.*
2. *There exists a bounded domain $\Omega_1 \supset \overline{\Omega}$ such that $g(x,y)$ can be extended to a C^∞ function on $\{(x,y) \in \Omega_1 \times \Omega_1 | \, x \neq y\}$.*
3. *Within a sufficiently small neighbourhood of each point $y \in \overline{\Omega}$ there holds $g(x,y) \in \Psi(\geq \nu, y)$.*
4. *The last condition remains true if the arguments x and y are exchanged. This means, to each fixed $x \in \overline{\Omega}$ there exists a constant $\varepsilon(x) > 0$ such that for all y with $0 < |x - y| < \varepsilon(x)$ the function $g(x,y)$ possesses the convergent series expansion with pseudohomogeneous terms*

$$g(x,y) = \sum_{j=\nu}^{\infty} g_j^*(y - x), \qquad g_j^* \in \Psi(j, 0).$$

Clearly, the coefficients of the pseudohomogeneous functions g_j^* may depend on x. (For simplicity, this is not included in the notation.)

Lemma 2.9.1. *If all coefficients of the system (2.1) are real-analytic in $\overline{\Omega}$ and if the ellipticity condition (2.6) is fulfilled in $\overline{\Omega}$, then (2.19) can be replaced by the relation*

$$L_{jk}^{(\mu)}(x,y) \in \Psi\overline{\Omega}(s_k + t_j - n) \tag{2.36}$$

and (2.26) can be replaced by

$$R_{jk}^{(\mu)}(x,y) \in \Psi\overline{\Omega}(\mu + 1 + s_k - s_j - n). \tag{2.37}$$

Proof. For all fixed points y, each entry of the matrices generated by the iteration sceme (2.14), (2.15) is obviously real-analytic with respect to x in the domain $\Omega_1 \setminus \{y\}$. Moreover, (2.25) involves the assertion of the local convergent series expansions.

Vice versa, let the point $x \in \overline{\Omega}$ be fixed. Then the matrix entries are real-analytic functions with respect to y in the domain $\Omega_1 \setminus \{x\}$. This can be deduced from (2.14), (2.15) by the following arguments:
1. The coefficients of the operator $T = A_0(y,\partial) - A(x,\partial)$ are analytic with respect to y.
2. The coefficients of the cofactor matrix operator, $B(y,\partial)$, are analytic with respect to y.
3. The right inverse $H^{(-1)}(y,\partial)$ is determined by (1.56). The function $1/H(\xi)$ as well as the numbers $H_{\alpha\beta}^{\sharp}$ used there, depend analytically on y. (Here we need the ellipticity condition!)

It remains to show that a local series expansion with pseudohomogeneous terms is possible for fixed $x \in \overline{\Omega}$. For this purpose, the iteration scheme (2.14), (2.15) is slightly modified. The symbol ∂ (= partial derivative with respect to x) is replaced by ∂_ϑ (= partial derivative with respect to ϑ). Then, in analogy to Lemma 1.13.2 step (iii), we can show that $H^{(-1)}(x,\partial_\vartheta)\delta(\vartheta)$ is a pseudohomogeneous function of degree $2\kappa - n$ with respect to ϑ, whereby the coefficients depend on x; and the parity condition is fulfilled.

The representation (2.12) is re-arranged as follows:

$$A_0^{(-1)} = B(x - \vartheta, \partial_\vartheta) \left(\sum_{\nu=0}^{\infty} \Theta^\nu \right) H^{(-1)}(x, \partial_\vartheta), \tag{2.38}$$

$$\Theta := H^{(-1)}(x,\partial_\vartheta)[H(x,\partial_\vartheta) - H(y,\partial_\vartheta)], \tag{2.39}$$

$$H(y,\partial_\vartheta) = H(x,\partial_\vartheta)(I - \Theta).$$

Now we choose a sufficiently small constant $\varepsilon = \varepsilon(x)$ such that, firstly, the Taylor series of all coefficients of the system $A(x,\partial)$ with midpoint x have a radius of convergence $\geq \varepsilon$ and such that, secondly, the operator (2.39) performs a contraction in the topological space $S'(\mathbb{R}^n)$ for each point y from the ε-neighbourhood of x.

Within the representation of the operators Θ and T all coefficients depending on y can then be replaced locally by Taylor expansions about x.

Formula (2.38) implies then firstly that the components of $N^{(0)}(x, y)$ admit a series expansion with pseudohomogeneous terms in the ε-neighbourhood of the point x. All terms have the same minimal degree of homogeneity as prescribed by (2.19). Secondly, the parity condition is fulfilled, since the differentiation with respect to ϑ and the multiplication with Taylor expansions in ϑ preserves the parity condition.

With the same arguments, the corresponding assertions can be obtained succesively for the matrices $N^{(\mu)}(x, y), L^{(\mu)}(x, y)$ and $R^{(\mu)}(x, y)$.   ∎

2.10 Classical Methods to Construct Fundamental Solutions for Elliptic Systems

2.10.1 Systems with Constant Coefficients

In the case of constant coefficients, the following universal method can be applied, which is sometimes called "Hörmander's cofactor method". (But it should not be confused with "Hörmander's method of descent"!)

A scalar partial differential equation with constant coefficients can be obtained by multiplying the complete operator with the cofactor matrix, anlogously to the procedure described in section (2.4). The construction of a fundamental solution for such scalar equations can be reduced to an ordinary differential equation by the "plane wave decomposition", developed by F. John [73]. If this ordinary differential equation cannot be solved in closed form, it is at least possible to construct a global convergent series expansion for the solution by use of the Fourier transform. Insertion and application of the cofactor matrix gives a fundamental solution for the original system.

2.10.2 J. Hadamard's Method

A method to construct fundamental solutions for equations with variable coefficients was originally described by J. Hadamard [55] for the special case of second order scalar equations with analytic coefficients. The first step in Hadamard's method consists in the derivation of a bilinear form from the second order terms and the introduction of a geodesic line (with respect to the metric generated by this bilinear form) which connects the points x and y. This is equivalent to the problem of finding such a change of variables which generates constant coefficients by the positions of the highest derivatives. Thereafter, a series expansion of the fundamental solution is derived. The consecutive terms are determined by a recursion equation which contains a path integral along the geodesic line.

The basic idea of Hadamard, to generate a series expansion of the fundamental solution by recursive application of lower order parts of the operator and a right inverse to the principal part is also used in our method described in section 2.4. In the contrary to Hadamard's method, an additional change of variables is not necessary.

2.10.3 Picard's Method of Iterated Kernels

Originally this method was developed for scalar equations. A generalization to the case of systems can be formulated as follows. The starting point of the iteration is the same as in our sceme, namely the matrix $L^{(0)}(x, y)$ determined by (2.14). The step (2.15) is replaced by

$$L^{(\iota+1)}(x, y) = L^{(\iota)}(x, y) + \int_{\Omega} L^{(\iota)}(x, t) R^{(\iota)}(t, y) dt, \qquad \iota = 0, 1, \dots \quad (2.40)$$

The remainders $R^{(\iota)}$ are derived from the actual $L^{(\iota)}$ according to formula (2.20). An application of the operator $A(x, \partial)$ to (2.40) leads to

$$\delta(x - y)I - R^{(\iota+1)}(x, y) = \delta(x - y)I - \int_{\Omega} R^{(\iota)}(x, t) R^{(\iota)}(t, y) dt,$$

which gives an iteration sceme for the components of the remainders, namely:

$$R_{jk}^{(\iota+1)}(x, y) = \sum_{i=1}^{n} \int_{\Omega} R_{ji}^{(\iota)}(x, t) R_{ik}^{(\iota)}(t, y) dt.$$

From formula (2.23) we obtain

$$R_{ji}^{(0)}(x, t) \in \Psi(\geq 1 + s_i - s_j - n, t); \qquad R_{ik}^{(0)}(t, y) \in \Psi(\geq 1 + s_k - s_i - n, y).$$

Proposition 2.10.1. *Let the kernels $K(x, y)$ and $K'(x, y)$ be pseudohomogeneous with respect to $x - y$ of degrees $\mu - n$ and $\nu - n$, respectively. Then the integral*

$$\int_{\Omega} K(x, t) K'(t, y) dt \qquad (2.41)$$

behaves like $|x - y|^{\mu + \nu - n}$ for $|x - y| \to 0$.

In the case $\mu, \nu > 0$ this is a classical result by G. Giraud (see [89, assertion 11.I]). It can be carried over to the hypersingular case if the integral (2.41) is understood in the partie finie sense.

Proposition 2.10.1 leads to

$$R_{jk}^{(1)}(x, y) \in \Psi(\geq 2 + s_k - s_j - n, y)$$

and step by step to

$$R_{jk}^{(\iota)}(x, y) \in \Psi(\geq 2^{\iota} + s_k - s_j - n, y).$$

As can be seen, Picard's iteration generates a sequence of Levi functions with an exponential growth of the degrees.

2.10.4 F. John's Method

An other method for the construction of fundamental solutions was developed by F. John [72], [73]. At first it requires the solution of special Cauchy problems for the differential operator considered. This has to be done for all hyperplanes containing the actual point x. Afterwards, the solutions must be multiplied with suitable weight functions and integrated with respect to all directions of the hyperplanes.

This method provides local existence results and gives insight into the structure of the fundamental solution in the vicinity of the singularity.

2.10.5 The Method of Freezing the Coefficients

Sometimes it is also called Schauder's method or Korn's trick. A description can be found e. g. in [63]. As explained in the first subsection, for the operator with frozen coefficients, namely $A(y, \partial)$, it is (in principle) always possible to construct a one-sided inverse. Now, the difference $A(x, \partial) - A(y, \partial)$ is a small pertubation for sufficiently small distances $|x-y|$ and the Neumann series can be applied to achieve a local convergent series expansion of the fundamental solution.

2.10.6 Comparison of the Methods

All the classical methods described above are constructive, however not suitable for the practical calculation of fundamental solutions in the case of variable coefficients. No such method is known - except for some very special cases [19],[20].

Each of the classical methods desribed above can be applied in principle to construct Levi functions of arbitrary degree by means of cutting the corresponding series expansions. Compared with the method from section (2.4), however, the computational time could be significantly higher because at least two processes are always nested and at least one of them can be realized by numerical approximations only.

Hadamard's method requires the determination of the geodesic connection for all pairs of points (x, y), which in general is only realizable by use of series expansions. Picard's method of iterated kernels is also very time consuming since the integrals in (2.40) are in general not computable in closed form. Instead, numerical quadrature is necessary here. Within the method of frozen coefficients, the inversion of the operator $A(y, \partial)$ can be realized only via series expansions. An additional problem is the continuation of the local expansions. The solution of the Cauchy problem *for all* hyperplanes (in discretized version for a large number) causes the main expense in F. John's method.

The method described here (in sect. 2.4) can be realized in each step by analytical calculations or comparison of coefficients, at least for $n \leq 2$. It turns out to be a combination of Hadamard's method and the method of frozen coefficients. It is, however, essentially less time consuming than either of both methods. It proves to be enough to freeze the coefficiets at the highest derivatives. Then the right inverse must be constructed only for an homogeneous differential operator with constant coefficients. This is much easier than for an operator which contains lower order terms, too.

3. Systems of Integral Equations, Generated by Levi Functions

First, we will sketch a well-known method to transform boundary value problems into systems of integral equations by use of Levi functions. This method goes back to E. E. Levi [81] and D. Hilbert [59, 60]. A more detailed description can be found e. g. in [89, chapter III] and the papers refered there. Moreover, this literature contains many results about the solvability conditions of the arising integral equations in the spaces $C^{k,\lambda}$ (= spaces of k times Hölder-continuously differentiable functions). Thereby, the boundary of the considered domain is assumed to be sufficiently smooth.

The theory worked out there does not contain the application we are interested in, namely the shell equations in polygonal domains. Due to the corners of the domain, singularities arise which require a special analysis. Sobolev spaces are more suitable for this purposes. Sections 3.3 and 3.4 collect some definitions and basic properties of Sobolev spaces which are of interest within the framework considered here. This material is well-known in essential.

Probably new, however, are some of the results obtained in the following sections. Here the mapping properties of the four basic types of integral operators are studied, namely operators acting from Sovolev spaces on Γ or Ω into Sobolev spaces on Γ or Ω. The boundary $\Gamma = \partial\Omega$ is assumed to be a curvilinear polygonal and the Mellin transformation technique is used for detailed investigations of the mapping properties. Up to now such investigations where only performed for integral (or pseudodifferential) operators acting from Γ to Γ (see [23] and the references there).

The results which we obtain for the different types of operators are used later in chapter 6 to analyze the system of integral equations derived from the shell equations. Some of the results are formulated more generaly here than needed later. So, they are also applicable within the framework of a more general theory. For instance, the polygonal boundaries fall into the general class of Lipschitz boundaries and some results are formulated just for this case.

3.1 Transformation of Boundary Value Problems into Systems of Integral Equations

Let $A(x, \partial)$ be an elliptic $m \times m$ system of partial differential operators of order 2κ. Together with appropriate boundary conditions, this leads to a boundary value problem which can be written in the general form

$$
\begin{aligned}
A(x, \partial)v(x) &= -p(x), \quad \forall x \in \Omega, \\
\mathbf{tr}_x \, S(x, \partial)v(x) &= \phi(x), \quad \forall x \in \Gamma.
\end{aligned} \tag{3.1}
$$

Here, $\mathbf{tr}_x = \mathbf{tr}_{x \to \Gamma}$ is the trace operator and $S(x, \partial)$ is a $\kappa \times m$ matrix of differential operators.

Subsequently we use the abbreviations $A_x := A(x, \partial)$ and $S_x := S(x, \partial)$. Sometimes x will be replaced by y. For instance, the operator $S_y := S(y, \partial_y)$ contains partial derivatives with recpect to y; $\mathbf{tr}_y$ means the trace $y \to \Gamma$. If $\mathbf{tr}$ stands alone, the trace with respect to x is meant.

We suppose that (3.1) is an *regular elliptic boundary value problem*. We don't want to give the definition here, since there exists a lot of literature on this topic where one can also find detailed investigations of solvability properties of such problems, including regularity results (especially in the vicinity of non-smooth boundary points). Per example we mention the monograph [90].

By use of a Levi function, boundary value problems can be transformed into coupled systems of boundary and domain integral equations. Thereby one can choose among various possibilities (like for the derivation of boundary integral equations in case that a fundamental solution is available).

3.1.1 The Direct Method

In analogy to the direct BEM one starts from the second Green's formula

$$
\int_\Omega \left\{ [v(x)]^\top A_x^* u(x) - [u(x)]^\top A_x v(x) \right\} dx =
$$
$$
= \int_\Gamma \left\{ [S_x v(x)]^\top \left[\tilde{T}_x u(x) \right] - [T_x v(x)]^\top \left[\tilde{S}_x u(x) \right] \right\} d\Gamma_x.
$$

Hereby $T_x, \tilde{S}_x, \tilde{T}_x$ is a corresponding (not uniquely determined) supplementary set of $\kappa \times m$ operator matrices. If the column vector $u(x)$ is replaced by the Levi-Funktion $L^*(x, y)$ of the adjoint (!) operator A_x^* then we obtain (after transposition and re-arrangement):

$$
v(y) - \int_\Omega [R^*(x, y)]^\top v(x) dx + \int_\Gamma [\tilde{S}_x L^*(x, y)]^\top [T_x v(x)] d\Gamma_x =
$$
$$
= \int_\Omega [L^*(x, y)]^\top p(x) dx + \int_\Gamma \left[\tilde{T}_x L^*(x, y) \right]^\top \phi(x) d\Gamma_x.
$$

This system of m domain integral equations of the second kind, together with the system of κ boundary integral equations obtained by application of the operator $\mathbf{tr}_y B_y$, forms a coupled system of integral equations for the unknowns $v(x)$ and the supplementary boundary data $\mathbf{tr}\, T_x v(x)$.

In this context we refer to the paper [25] where the concept of strong ellipticity is worked out for elliptic boundary value problems and where the question is investigated, under which conditions the strong ellipticity carries over to the obtained boundary integral equations. Further contributions to this topic can be found in [135].

3.1.2 The Indirect Method

From BEM it is well-known that the solution of a boundary value problem can be represented only by a simple layer potential or only by a double layer potential. This is the well-known *indirect method*. Its mathematical theory was worked out by G. Fichera and P. Ricci [39, 103].

In analogy to the simple layer approach, the solution is sought now in the form

$$v(x) = \int_\Omega L^\Omega(x,y)w(y)dy + \int_\Gamma L^\Gamma(x,y)f(y)d\Gamma_y. \tag{3.2}$$

Hereby, $L^\Omega(x,y)$ is a Levi function, i. e., a $m \times m$ matrix fulfilling the equation

$$A_x L^\Omega(x,y) = \delta(x-y)I_{m \times m} - R^\Omega(x,y). \tag{3.3}$$

The remainder $R^\Omega(x,y)$ contains only pseudohomogeneous terms of degree $\geq 1 - n$. The other kernel in (3.2) is the $m \times \kappa$ matrix

$$L^\Gamma(x,y) := \left\{ S_y \left[L^\Omega(x,y) \right]^\top \right\}^\top. \tag{3.4}$$

If $\phi(x)$ has the meaning of Dirichlet data, then the boundary integral in (3.2) can be interpreted as simple layer potential.

An application of the $m \times m$ operator matrix A_x to the ansatz (3.2) generates the following system of m domain integral equations:

$$w(x) - \int_\Omega R^\Omega(x,y)w(y)d\Omega_y - \int_\Gamma R^\Gamma(x,y)f(y)d\Gamma_y = -p(x), \quad \forall x \in \Omega, \tag{3.5}$$

with

$$R^\Gamma(x,y) = -A_x L^\Gamma(x,y) = \left\{ S_y \left[R^\Omega(x,y) \right]^\top \right\}^\top. \tag{3.6}$$

The application of $\mathbf{tr}_x S_x$ to (3.2) leads to a system of κ boundary integral equations:

$$\int_\Omega V^\Omega(x,y)w(y)d\Omega_y + \int_\Gamma V^\Gamma(x,y)f(y)d\Gamma_y = \phi(x), \qquad \forall x \in \Gamma. \tag{3.7}$$

V^Ω denotes the $\kappa \times m$ matrix

$$V^{\Omega}(x,y) := S_x L^{\Omega}(x,y) \tag{3.8}$$

and V^{Γ} is the $\kappa \times \kappa$ matrix

$$V^{\Gamma}(x,y) := S_x L^{\Gamma}(x,y). \tag{3.9}$$

In short notation, the system (3.5), (3.7) can be written as:

$$\begin{pmatrix} I - R^{\Omega} & -R^{\Gamma} \\ V^{\Omega} & V^{\Gamma} \end{pmatrix} \begin{pmatrix} w \\ f \end{pmatrix} = \begin{pmatrix} -p \\ \phi \end{pmatrix}. \tag{3.10}$$

For simplicity, the integral operators are denoted by the same letters like the corresponding kernels.

3.1.3 Comparison of the Methods

Compared with the indirect method, the direct method has the advantage that the solution[1] of the integral equation system has no corner singularities if the boundary data admits a smooth solution of the differential equations. The last case is sometimes of interest, especially, if the boundary value problem comes from a domain decomposition method and if Ω is an interior subdomain.

The disadvantage of the direct method consists in a higher code developing time, since more integral kernels must be calculated and discretized. This fact is obvious, if only the boundary value problem (3.1) is to be solved. Often one is also interested in the complementary boundary data (e. g., in the boundary traction if the boundary displacements is pre-determined and vice versa). The direct method produces them immediately, the indirect method requires a post-processing step (like in BEM). Also in this case, however, the indirect method leads to fewer code developing expenses, at least, if one measures the length of the code. Applying the indirect method to the shell equations, the 3×3 matrix (6.22) can be used as a Levi function. It has 5 non-vanishing entries, a suitable Levi function for the direct method has 7 non-vanishing entries. All together, the indirect method requires the computation and discretization of 66 integral kernels (inclusive post-processing); for the direct method we have 105 kernels.

Consequently, the indirect method leads to fewer expenses for just writing of the code. It is well-known, however, that the main part of the code developing time is consumed by tests and debugging. From this viewpoint again, the direct method appears to be of advantage, because the solution of the integral equation system can be tested immediately with respect to their (approximate) correctness. Altogether, the decision is not unique, for favouring one of both methods.

Since we have used the indirect method for our numerical tests, we will confine ourselves to it in what follows.

---

[1] This assertion concerns only the exact solution. Due to the approximation error, the numerical solution may contain certain projections of the corresponding corner singularities (multiplied with a small factor) onto the trial space.

3.2 On the Unique Solvability of the Integral Equation System

3.2.1 Sufficient Conditions

For simplification we avoid in the following lemma specifications, in which spaces the operators are considered. We only assume "appropriate spaces".

Lemma 3.2.1. *Assume that the following three conditions are satisfied:*

1. *The boundary value problem (3.1) is uniquely solvable.*
2. *$L^{\Omega}(x,y)$ is the fundamental solution of some (arbitrary) differential operator A_0.*
3. *The boundary integral operator V^{Γ} has a trivial zero-space.*

Then the system (3.10) with a vanishing right-hand side $p(x) = \phi(x) = 0$ has only the trivial solution $w(x) = f(x) = 0$.

Proof. Vanishing of $p(x)$ and $\phi(x)$ has the following consequences for the function (3.2): $A_x v(x) = 0$ in Ω and $S_x v(x) = 0$ on Γ. Thus, the first assumption implies $v = 0$ in Ω.

The application of the operator A_0 to $L^{\Omega}(x,y)$ produces the δ-distribution; the application to $L^{\Gamma}(x,y)$ generates a certain distribution with a support on Γ. Hence, the application of A_0 to (3.2) gives immediately: $w(x) = v(x) = 0$.

Only the boundary integral remains. From the third assumption then follows $f(x) = 0$. ∎

The first assumption of Lemma 3.2.1 is already contained in the requirement of ellipticity for the boundary value problem. The third assumption is not an essential restriction since for the boundary integral operators there holds Fredholm's alternative with index 0 in appropriate spaces. If the dimension D_K of the zero-space is greater than zero, invertibility in an extended space can be achieved by introducing D_K additional unknowns and adding D_K suitable functionals to the system. This concept will be applied later for the shell equations (see section 6.3).

The second condition is obvious as well, provided that the used Levi function contains only the principal singularity, i. e., only the term $L^{(0)}(x,y) = N^{(0)}(x,y)$ determined by (2.14). If one takes further terms in order to obtain higher smoothness for the remainder, the second assumption and the uniqueness may be violated as can be seen from the following example.

3.2.2 An Example for Non-Unique Solvability

We consider again the one-dimensional differential operator (2.31) in an interval (Y_1, Y_2) with $0 < Y_1 < Y_2$. The coefficients are choosen as: $a_2(x) = x$, $a_1(x) = 1$, $a_0(x) = 0$. We use the Levi function (2.32) of second degree, which now takes the form

$$L(x,y) = P[< 3] \, L^{(1)}(x,y) = \frac{|x-y|}{2y}\left(1 - \frac{x-y}{2y}\right).$$

Then the equation

$$v(x) = \int_{Y_1}^{Y_2} L(x,y)w(y)dy + f_1 L(x,Y_1) + f_2 L(x,Y_2) \equiv 0$$

has in the case $Y_2 = (2+\sqrt{3})Y_1$ (and only in this case) a non-trivial solution, namely:

$$w(y) = y^2, \quad f_1 = (2+\sqrt{3})Y_1^3, \quad f_2 = -(7+4\sqrt{3})Y_1^3.$$

3.3 Sobolev Spaces

3.3.1 Spaces on $\mathbb{R}^n$

For real numbers s, the Sobolev space $H^s(\mathbb{R}^n)$ is defined as the set of all distributions with a local integrable Fourier transform $\hat{f}(\xi)$ and a finite norm [35, formula 4.1]

$$\|f; H^s(\mathbb{R}^n)\| := \left[\int_{\mathbb{R}^n} \left|\hat{f}(\xi)\right|^2 (1+|\xi|)^{2s}\, d\xi\right]^{\frac{1}{2}}. \tag{3.11}$$

If s is not an integer, these spaces are also called Sobolev-Slobodeckiĭ spaces.

From (3.11) one can see that the δ-distribution belongs to $H^s(\mathbb{R}^n)$ for $s < -n/2$. For $s > n/2$, all functions from H^s are continuous [50, Theorem 1.4.4.1]. Discontinuous and unbounded functions are contained in H^s already for $s = n/2$ [50, sect. 1.1].

The subset of all distributions from $H^s(\mathbb{R}^n)$ with compact support is denoted by $H^s_{\text{comp}}(\mathbb{R}^n)$. Hence, it is in a natural way the dual space to $H^{-s}_{\text{loc}}(\mathbb{R}^n)$, the space of all distributions belonging locally (i. e., after multiplication with a suitable cut-off function) to H^{-s}.

For instance, the following functions belong to the spaces H^s_{loc}:

$$|x|^t (\ln|x|)^l \in H^s_{\text{loc}}(\mathbb{R}^n), \qquad \forall l \in \mathbb{N}_0, \, \forall s, t \in \mathbb{R}^1 \quad \text{with} \quad s < t + n/2. \tag{3.12}$$

This can be deduced from the properties of the Fourier transform (see Lemma 1.3.2).

We still establish the following embedding theorem:

Lemma 3.3.1. Let $0 \le t \le 1$. Then

$$\int_{\mathbb{R}^n} |x|^{2t} |grad\, u(x)|^2 dx < \infty \Longrightarrow u \in H^{1-t}_{\text{loc}}(\mathbb{R}^n).$$

Proof. For $t = 0$ the implication is trivial and the case $t = 1$ is contained in the more result [77, Lemma 4.9]. The intermediate values can be treated by interpolation. ∎

3.3.2 Spaces on Lipschitz Domains

Suppose $\Omega \subset \mathbb{R}^n$ is a Lipschitz domain, i. e., an open set with a Lipschitz continuous boundary (compare [50, Def. 1.2.1.2]). The space $H^s(\Omega)$ is defined for $s \geq 0$ as the set of all functions $f(x)$, which are square integrable over Ω and have a finite H^s norm, defined as follows (see [50, sect. 1.3]):

– for $s \in \mathbb{N}_0$ is

$$\|f(x); H^s(\Omega)\|^2 := \sum_{|\beta|=0}^{s} \int_\Omega |\partial^\beta f(x)|^2 \, dx; \tag{3.13}$$

– for $\sigma \in (0,1)$, $m \in \mathbb{N}_0$ and $s = m + \sigma$ is

$$\|f(x); H^s(\Omega)\|^2 := \|f(x); H^m(\Omega)\|^2 +$$
$$+ \sum_{|\beta|=m} \int_\Omega \int_\Omega \frac{|\partial^\beta f(x) - \partial^\beta(x')|^2}{|x - x'|^{n+2\sigma}} \, dx \, dx'.$$

Alternatively, $H^s(\Omega)$ can be introduced for $s \geq 0$ as the restriction of all functions from $H^s(\mathbb{R}^n)$ to Ω [50, Def. 1.3.2.4 and Theorem 1.4.3.1]. Both definitions are equivalent and we have the following norm estimate:

$$\underline{C}_{s,\Omega} \|f(x); H^s(\Omega)\| \leq \inf_{F|_\Omega = f} \|F(x); H^s(\mathbb{R}^n)\| \leq \overline{C}_{s,\Omega} \|f(x); H^s(\Omega)\|.$$
$$\tag{3.14}$$

For $s > 0$, we denote by $\overset{\circ}{H}{}^s(\Omega)$ the closure of

$$\mathcal{D}(\Omega) := \{u(x) \in C_0^\infty(\mathbb{R}^n) \,|: \, \operatorname{supp} u \subset \Omega\}$$

in the H^s norm [50, Def. 1.3.2.2]; and $H^{-s}(\Omega)$ denotes the dual space to $\overset{\circ}{H}{}^s(\Omega)$ with respect to the L^2 scalar product [50, Def. 1.3.2.3].

We still need the spaces $\widetilde{H}^s(\Omega)$. For $s \geq 0$, the space $\widetilde{H}^s(\Omega)$ is the set of all functions $u(x)$ defined on Ω, for which the "continuation by zero", denoted by $u^\sim$, belongs to $H^s(\mathbb{R}^n)$ globally. $\widetilde{H}^{-s}(\Omega)$ is the dual space to $H^s(\Omega)$ with respect to the scalar product in L^2.

Let χ_Ω denote the characteristic function of the domain Ω. For $s > 0$ and $u \in H^{-s}(\Omega)$, the continuation by zero $,u^\sim$, is defined by

$$< u^\sim, \phi >\, := \,< u, \phi \chi_\Omega >, \qquad \forall \phi \in H^s(\mathbb{R}^n).$$

This is a distribution from $H^{-s}(\mathbb{R}^n)$ with support in $\overline{\Omega}$. With the constants from (3.14) there hold the estimates:

$$|< u^\sim, \phi >| \;\leq\; \left\| u; \widetilde{H}^{-s}(\Omega) \right\| \cdot \| \phi \chi_\omega; H^s(\Omega) \|$$
$$\leq\; \underline{C}_{s,\Omega}^{-1} \left\| u; \widetilde{H}^{-s}(\Omega) \right\| \cdot \| \phi; H^s(\mathbb{R}^n) \|,$$
$$\| u^\sim; H^{-s}(\mathbb{R}^n) \| \;\leq\; \underline{C}_{s,\Omega}^{-1} \left\| u; \widetilde{H}^{-s}(\Omega) \right\|. \tag{3.15}$$

Lemma 3.3.2. *For $-1/2 < s < 1/2$, the spaces $H^s(\Omega)$ and $\tilde{H}^s(\Omega)$ coincide.*

Proof. The coincidence of the spaces $H^s(\Omega)$, $\tilde{H}^s(\Omega)$ and $\overset{\circ}{H}{}^s(\Omega)$ for $0 \leq s < 1/2$ is verified in [50, Corollary 1.4.4.5]. The coincidence of the dual spaces follows then immediately. ∎

3.3.3 Spaces on the Boundary of Lipschitz Domains

For $s = 0$, set $H^0(\Gamma) = L^2(\Gamma)$ and for $s > 0$:

$$H^s(\Gamma) \;\; := \;\; \left\{ \operatorname{tr} u \,\middle|\, u \in H^{s+1/2}(\mathbb{R}^n) \right\},$$

$$H^{-s}(\Gamma) \;\; := \;\; \text{dual space to } H^s(\Gamma) \text{ with respect to the } L^2 \text{scalar product.}$$

If Γ is a polygon in the plane which consists of the arcs $\Gamma_1, \ldots, \Gamma_J$, then the following representation holds for $s \geq 0$:

$$H^s(\Gamma) = \prod_j H^s(\Gamma_j). \tag{3.16}$$

For $s \geq 1/2$, some additional compatibility conditions must be satisfied (see [50, sect. 1.5.2]).

3.3.4 Spaces on the Cut Plane

Wedge-shaped domains in the plane are described by polar coordinates as follows:

$$\mathbb{K}[\varphi_1, \varphi_2] \;\; := \;\; \left\{ x = (r \cos \varphi, r \sin \varphi) \,\middle|\, : \; 0 < r < \infty, \; \varphi_1 < \varphi < \varphi_2 \right\},$$
$$\text{with} \;\; 0 \leq \varphi_1 < \varphi_2 \leq 2\pi,$$
$$\mathbb{K} \;\; := \;\; \mathbb{K}[0, 2\pi].$$

For $s \geq 0$, the space $H^s(\mathbb{K})$ is defined by the norm

$$\|v; \, H^s(\mathbb{K})\| \;\; := \;\; \lim_{\varepsilon \to 0} \|v; \, H^s(\mathbb{K}[\varepsilon, 2\pi - \varepsilon])\|. \tag{3.17}$$

Lemma 3.3.2 implies that the spaces $H^s(\mathbb{K})$ and $H^s(\mathbb{R}^2)$ coincide for $0 \leq s < 1/2$. This is not true anymore for $s > 1/2$. The functions from $H^s(\mathbb{K})$ are allowed to be discontinuous on the positive half-axis.

3.4 Mellin Transform and Weighted Sobolev Spaces

The Mellin transform of a function $f(x) \in C_0^\infty(\mathbb{R}_+)$ is defined by

$$\mathcal{M}[f](\lambda) := \int_0^\infty x^{i\lambda-1} f(x) dx, \qquad \forall \lambda \in \mathbb{C}.$$

The following basic properties can easily be proved:

$$\mathcal{M}[x^\gamma f(x)](\lambda) = \mathcal{M}[f](\lambda - i\gamma), \qquad \forall \gamma \in \mathbb{R}^1,$$

$$\mathcal{M}[f'(x)](\lambda) = (1 - i\lambda)\mathcal{M}[f](\lambda + i),$$

$$\int_0^\infty x^{2\gamma-1} |f(x)|^2 dx = \frac{1}{2\pi} \int_{\Im m\, \lambda = -\gamma} |\mathcal{M}[f](\lambda)|^2 d\lambda, \qquad \forall \gamma \in \mathbb{R}^1.$$

The last relation is called *Parseval's identity* for the Mellin transform.

The Mellin transform can be extended to larger classes of functions and distributions. In general, the image is then no longer defined for all $\lambda \in \mathbb{C}$, but only on a certain subset, the *range of definition*. In many cases the Mellin transform possesses a meromorphic continuation to the whole complex plane. If a Mellin transform is written as a meromorphic function, the original range of definition must always be declared. In the cases considered here, the ranges of definition are strips parallel to the real axis.

In the following Definition 3.4.1, some norms are introduced by integrating the Mellin transform of certain distributions along lines parallel to the real axis. This notation involves the assumption that the integration line belongs to the range of definition for the corresponding Mellin transform. Taking this onto account, one sees, that the spaces introduced in Definition 3.4.1 could be equivalently defined as the closure (in the corresponding norm) of the set of those functions, which are a C_0^∞ functions with respect to $r = |x|$. This access is used e. g. in [7, 22, 27, 77] to introduce weighted Sobolev spaces.

Definition 3.4.1. *For $s, \gamma \in \mathbb{R}^1$, the space $\overset{\circ}{W}{}_\gamma^s(\mathbb{R}_+)$ is defined as the set of all distributions $f \in \mathcal{S}'(\mathbb{R}_+)$ with a finite norm*

$$\left\| f; \overset{\circ}{W}{}_\gamma^s(\mathbb{R}_+) \right\|^2 := \frac{1}{2\pi} \int_{\Im m\, \lambda = s - \frac{\gamma+1}{2}} (1 + |\lambda|^2)^s \, |\mathcal{M}[f](\lambda)|^2 \, d\lambda. \qquad (3.18)$$

For $\gamma \in \mathbb{R}^1$, $m \in \mathbb{N}_0$, $\sigma \in [0,1)$ and $s = m + \sigma$, the space $\overset{\circ}{W}{}_\gamma^s(\mathbb{K})$ is defined as the set of all distributions $u(r, \phi) \in \mathcal{S}'(\mathbb{R}^2)$ with finite norm

$$\left\| u(r, \phi); \overset{\circ}{W}{}_\gamma^s(\mathbb{K}) \right\|^2 := \qquad\qquad\qquad\qquad\qquad\qquad\qquad (3.19)$$

$$:= \frac{1}{2\pi} \lim_{\varepsilon \to 0} \sum_{j=0}^m \int_{\Im m\, \lambda = s - 1 - \gamma/2} \int_\varepsilon^{2\pi-\varepsilon} (1 + |\lambda|^2)^{m-j} \left| \frac{\partial^j \mathcal{M}[u](\lambda, \phi)}{\partial \phi^j} \right|^2 d\phi \, d\lambda.$$

Hereby, $\mathcal{M}[u](\lambda, \phi)$ denotes the Mellin transform of $u(r, \phi)$ in radial direction.

Lemma 3.4.1. *Let $\Omega = \mathbb{R}_+$ or $\Omega = \mathbb{K}[a,b]$. In both cases we have the following properties:*

(i)   *The multiplication with $|x|^t$ is for all $t \in \mathbb{R}^1$ a bijective mapping from $\overset{\circ}{W}{}^s_\gamma(\Omega)$ to $\overset{\circ}{W}{}^s_{\gamma-2t}(\Omega)$.*

(ii) $\overset{\circ}{W}{}^s_\gamma(\Omega)$ *is continuously embedded into $\overset{\circ}{W}{}^{s-t}_{\gamma-2t}(\Omega)$, for all $t > 0$.*

(iii) *The operators of differentiation, namely d/dx in case n=1 and ∂_1, ∂_2 in*

   case n=2, are continuous mappings from $\overset{\circ}{W}{}^s_\gamma(\Omega)$ to $\overset{\circ}{W}{}^{s-1}_\gamma(\Omega)$.

(iv) $\overset{\circ}{W}{}^m_0(\Omega) = \overset{\circ}{H}{}^m_0(\Omega), \qquad \forall m \in \mathbb{N}_0$.

These propositions follow directly from the definition of the norms and the properties of the Mellin transform. Further, from Lemma 3.3.1 we obtain:

Lemma 3.4.2. *For $0 \le t \le 1$ there holds the implication*

$$\lim_{\varepsilon \to 0} \int_{\mathbb{K}[\varepsilon, 2\pi - \varepsilon]} |x|^{2t} \, |grad\, u(x)|^2 \, dx < \infty \implies u \in H^{1-t}_{\mathrm{loc}}(\mathbb{K}).$$

Proof. We choose a number $\tau \in (0, \pi)$ and denote the restrictions of $u(x)$ to $\mathbb{K}[0, \pi + \tau]$ and to $\mathbb{K}[\pi - \tau, 2\pi]$ by $u_+(x)$ and $u_-(x)$, respectively. Both can be extended to functions on $\mathbb{R}^2$ in such a way that Lemma 3.3.1 can be applied. If, afterwards, the functions are restricted again to the corresponding subdomains, we obtain:

$$\begin{aligned} u_+ &\in H^{1-t}_{\mathrm{loc}}(\mathbb{K}[0, \pi + \tau], \\ u_- &\in H^{1-t}_{\mathrm{loc}}(\mathbb{K}[\pi - \tau, 2\pi]). \end{aligned}$$

This immediately gives the assertion of the lemma.   ∎

3.5 Cauchy Singular Integral Operators on the Half Axis

Within this section we identify $\mathbb{R}^2$ with the complex plane $\mathbb{C}$. Accordingly, the pair of real numbers $x = (x_1, x_2) = (r \cos \phi, r \sin \phi)$ is identified with the complex number $z = x_1 + ix_2 = re^{i\phi}$.

3.5.1 A Basic Theorem

Let $G(z) = r^{-1}g(\phi)$ be a homogeneous function of degree -1. Contrary to the usual notation in function theory, $G(z)$ does not denote a holomorphic function. For the function $g(\phi)$ we require:

$$\begin{aligned} g(\phi + \pi) &= -g(\phi), && \forall \phi, & (3.20) \\ g(\phi) &\in L^p[0, 2\pi], && \forall p < \infty. & (3.21) \end{aligned}$$

(3.20) corresponds to the parity condition.

Theorem 3.5.1. *Under the assumptions* (3.20) *and* (3.21), *the integral operator*

$$(Gf)(z) := \int_0^\infty G(z - y)f(y)dy \qquad (3.22)$$

is for $0 \le s < 1$ and $|2s - \gamma| < 1$ a continuous mapping from $\overset{o}{W}{}^s_\gamma(\mathbb{R}_+)$ to $\overset{o}{W}{}^s_{\gamma-1}(\mathbb{K})$.

Proof. The Mellin transform of (3.22) in radial direction is

$$\mathcal{M}[Gf](\lambda, \phi) = \int_0^\infty r^{i\lambda-1}(Gf)\left(re^{i\phi}\right) dr.$$

We consider at first the special case $G(z) = \bar{z}^n \, z^{-(n+1)}$. (Within this proof, the letter n, which elsewhere denotes the space dimension, is used in other meaning.) By virtue of the binomial formula there holds:

$$(Gf)(re^{i\phi}) = \sum_{k=0}^n (-1)^{n-k} \binom{n}{k} e^{-ik\phi} \int_0^\infty \frac{r^k y^{n-k}}{(re^{i\phi} - y)^{n+1}} f(y)dy$$

$$= \sum_{k=0}^n (-1)^{1+k} \binom{n}{k} \exp[-i\phi(k + n + 1)] \int_0^\infty \frac{r^k y^{n-k}}{(ye^{-i\phi} - r)^{n+1}} f(y)dy.$$

The Mellin transform of such expressions is calculated in [22, Lemma 1]. For $\phi \in (0, 2\pi)$ and $\Im \lambda \in (k - n - 1, k)$ one finds:

$$\mathcal{M}\left[\int_0^\infty \frac{r^k y^{n-k}}{(ye^{i(2\pi-\phi)} - r)^{n+1}} f(y)dy\right](\lambda, \phi) =$$

$$= \mathcal{M}[f](\lambda) \frac{(-i)^{n+1}\pi}{n! \sinh(\pi\lambda)} \exp[i(n + 1 - k)\phi + (\phi - \pi)\lambda] \prod_{j=1}^n [\lambda + (j - k)i].$$

For $n = 0$, the product is equal to 1. Thus

$$\mathcal{M}[Gf](\lambda, \phi) = \frac{i\pi e^{(\phi-\pi)\lambda}}{\sinh(\pi\lambda)} \psi_n(\lambda, \phi) \mathcal{M}[f](\lambda), \qquad -1 < \Im \lambda < 0,$$

$$\psi_n(\lambda, \phi) := \sum_{k=0}^n \frac{(-1)^k}{n!} \binom{n}{k} e^{-2ik\phi} \prod_{j=1}^n (j - k - i\lambda)$$

$$= (-1)^n \sum_{k=0}^n \binom{i\lambda - 1}{n - k}\binom{-i\lambda}{k} e^{-2ik\phi}.$$

If ϕ is replaced by $2\pi - \phi$, we obtain for the case $G(z) = z^n \, \bar{z}^{-(n+1)}$ the Mellin transform

$$\mathcal{M}[Gf](\lambda, \phi) = \frac{i\pi e^{(\pi-\phi)\lambda}}{\sinh(\pi\lambda)} \, \psi_n(\lambda, -\phi) \, \mathcal{M}[f](\lambda) \qquad -1 < \Im \lambda < 0.$$

Now we consider to the general case $G(z) = r^{-1}g(\phi)$. Relation (3.20) implies that the Fourier series contains only odd exponents:

$$g(\phi) = \sum_{n=-\infty}^{\infty} g_n e^{i(2n-1)\phi}. \tag{3.23}$$

This implies

$$G(z) = \frac{g(\phi)}{r} = \sum_{n=0}^{\infty} \left(g_{n+1} \frac{z^n}{\bar{z}^{n+1}} + g_{-n} \frac{\bar{z}^n}{z^{n+1}} \right),$$

$$\mathcal{M}[Gf](\lambda, \phi) = \left[\widehat{G}_+(\lambda, \phi) + \widehat{G}_-(\lambda, \phi) \right] \mathcal{M}[f](\lambda) \tag{3.24}$$

$$(\text{for } -1 < \Im \lambda < 0),$$

$$\widehat{G}_+(\lambda, \phi) := \frac{i\pi}{\sinh(\pi\lambda)} e^{(\pi-\phi)\lambda} \sum_{n=0}^{\infty} g_{n+1}\psi_n(\lambda, -\phi), \tag{3.25}$$

$$\widehat{G}_-(\lambda, \phi) := \frac{i\pi}{\sinh(\pi\lambda)} e^{(\phi-\pi)\lambda} \sum_{n=0}^{\infty} g_{-n}\psi_n(\lambda, \phi). \tag{3.26}$$

It remains to investigate the behaviour of the functions (3.25), (3.26) when $|\Re \lambda| \to \infty$. An estimation from above by the sum of the absolute values of the series terms appears not to be sharp enough for this purposes. To achieve more precise estimates, we introduce the following product of two binomial series:

$$\Phi(\zeta, \lambda, \phi) := (1-\zeta)^{i\lambda-1} \left(1 - \zeta e^{-2i\phi} \right)^{-i\lambda}$$

$$= \sum_{m=0}^{\infty} \binom{i\lambda-1}{m} (-\zeta)^m \sum_{k=0}^{\infty} \binom{-i\lambda}{k} \left(-\zeta e^{-2i\phi} \right)^k.$$

These series converge absolutely for $\rho := |\zeta| < 1$. We set $n = m + k$ and re-arrange the double series as follows:

$$\Phi(\zeta, \lambda, \phi) = \sum_{n=0}^{\infty} \sum_{k=0}^{n} \binom{i\lambda-1}{n-k} \binom{-i\lambda}{k} (-\zeta)^n e^{-2ik\phi} = \sum_{n=0}^{\infty} \zeta^n \psi_n(\lambda, \phi).$$

This can be interpreted as a Fourier series in $\omega = \arg(\zeta)$. The representation formula for the Fourier coefficients gives

$$\frac{1}{2\pi} \int_0^{2\pi} \Phi(\rho e^{i\omega}, \lambda, \phi) e^{-in\omega} d\omega = \rho^n \psi_n(\lambda, \phi), \qquad 0 \le \rho < 1, \, n \in \mathbb{N}_0. \tag{3.27}$$

Multiplication of (3.27) with the Fourier coefficients of the functions

$$g_+(\phi) := \sum_{n=0}^{\infty} g_{n+1}e^{-in\phi}, \qquad g_-(\phi) := \sum_{n=0}^{\infty} g_{-n}e^{-in\phi} \qquad (3.28)$$

and adding up with respect to n gives the integral representations

$$\frac{1}{2\pi}\int_0^{2\pi} \Phi(\rho e^{i\omega}, \lambda, \phi)g_-(\omega)d\omega = \sum_{n=0}^{\infty} \rho^n g_{-n}\psi_n(\lambda, \phi), \qquad (3.29)$$

$$\frac{1}{2\pi}\int_0^{2\pi} \Phi(\rho e^{i\omega}, \lambda, -\phi)g_+(\omega)d\omega = \sum_{n=0}^{\infty} \rho^n g_{n+1}\psi_n(\lambda, -\phi). \qquad (3.30)$$

Condition (3.21) guarantees that both functions from (3.28) belong to the space $L^p[0, 2\pi]$ for all $p < \infty$. This can be verified from the continuity of the Cauchy-singular integral in these spaces and the well-known fact that the projections to the "half" Fourier series can be written as a linear combination of the identity and the Hilbert transform [98, sect. 3.4.4]. Hence it is ensured that the expressions on both sides of (3.29) and (3.30) are well-defined for $\rho < 1$.

Our next aim is the passing to the limit $\rho \to 1$. For this purpose it is first necessary to investigate the function $\Phi(\zeta, \lambda, \phi)$ more precisely. The range of definition can be desribed as follows:

$$\zeta = e^{i\omega}, \quad \omega \in [0, 2\pi); \quad \lambda = \sigma - i\theta, \quad \sigma \in (-\infty, \infty), \quad \theta \in (0, 1); \quad \phi \in (0, 2\pi).$$

To apply the complex logarithm, the plane is cut along the negative real axis. Therefore the arguments of all terms must be transformed to the interval $[-\pi, \pi)$. One of the occuring arguments is $\omega - \pi - 2\phi$. The transformed argument is written in the form $\omega - \tau - 2\eta$ with

$$\eta = \eta(\omega, \phi) = \begin{cases} \phi, & \text{for} \quad \phi < \pi, \ 2\phi \le \omega, \\ \phi - \pi, & \text{for} \quad 2(\phi - \pi) \le \omega < 2\phi, \\ \phi - 2\pi, & \text{for} \quad \pi < \phi, \ \omega < 2(\phi - \pi). \end{cases}$$

For $\zeta = e^{i\omega}$ we obtain:

$$\ln(1 - \zeta) = \ln\left(2\sin\frac{\omega}{2}\right) + i\frac{\omega - \pi}{2},$$

$$\ln\left(1 - \zeta e^{-2i\phi}\right) = \ln\left[2\sin\left(\frac{\omega}{2} - \eta\right)\right] + i\left(\frac{\omega - \pi}{2} - \eta\right),$$

$$\ln\Phi(\zeta, \lambda, \phi) = (\theta - 1 + i\sigma)\ln(1 - \zeta) - (\theta + i\sigma)\ln\left(1 - \zeta e^{-2i\phi}\right),$$

$$|\Phi(\zeta, \lambda, \phi)| = \exp(\Re \ln \Phi) = \Psi(\omega, \phi, \theta)e^{-\sigma\eta},$$

$$\Psi(\omega, \phi, \theta) := \left(2\sin\frac{\omega}{2}\right)^{\theta-1}\left[2\sin\left(\frac{\omega}{2} - \eta\right)\right]^{-\theta} =$$

$$= \frac{1}{2}\left|\sin\frac{\omega}{2}\right|^{\theta-1}\left|\sin\left(\frac{\omega}{2} - \phi\right)\right|^{-\theta}.$$

This leads to the following estimates for the functions (3.25), (3.26):

$$\left| \widehat{G}_+(\lambda, \phi) \right| \leq \frac{e^{(\pi - \phi)\sigma}}{2|\sinh(\pi\lambda)|} \max_\omega \exp[-\sigma\eta(\omega, 2\pi - \phi)] \cdot \tag{3.31}$$

$$\cdot \int_0^{2\pi} \Psi(\omega, \phi, \theta) \left| g_+(\omega) \right| d\omega,$$

$$\left| \widehat{G}_-(\lambda, \phi) \right| \leq \frac{e^{(\phi - \pi)\sigma}}{2|\sinh(\pi\lambda)|} \max_\omega \exp[-\sigma\eta(\omega, \phi)] \cdot \tag{3.32}$$

$$\cdot \int_0^{2\pi} \Psi(\omega, \phi, \theta) \left| g_-(\omega) \right| d\omega.$$

From

$$e^{(\phi - \pi)\sigma} \max_\omega \exp[-\sigma\eta(\omega, \phi)] = \begin{cases} e^{\pi|\sigma|}, & \text{for } (\pi - \phi)\sigma < 0, \\ 1, & \text{for } (\pi - \phi)\sigma \geq 0, \end{cases}$$

$$e^{(\pi - \phi)\sigma} \max_\omega \exp[-\sigma\eta(\omega, 2\pi - \phi)] = \begin{cases} 1, & \text{for } (\pi - \phi)\sigma \leq 0, \\ e^{\pi|\sigma|}, & \text{for } (\pi - \phi)\sigma > 0, \end{cases}$$

it is obvious that the complete expression in front of the integral on the right-hand side of (3.31) can be estimated by a constant C_θ depending only on θ. The same assertion holds for the corresponding term in (3.32).

In the sequel, we consider $\theta = -\Im \mathfrak{m} \lambda \in (0, 1)$ as fixed, choose a number q fulfilling

$$1 < q < \min\left(\frac{1}{\theta}, \frac{1}{1 - \theta}\right) \tag{3.33}$$

and denote by $p = q/(q - 1)$ the conjugate exponent. Then (3.32) can be estimated further by use of Hölder's inequality:

$$\left| \widehat{G}_-(\lambda, \phi) \right| \leq C_\theta \|g_-\|_{L^p} \left(\int_0^{2\pi} [\Psi(\omega, \phi, \theta)]^q d\omega \right)^{1/q} \tag{3.34}$$

$$= \frac{C_\theta}{2} \|g_-\|_{L^p} [P(\phi, \theta, q) + P(-\phi, \theta, q)]^{1/q}, \tag{3.35}$$

$$P(\phi, \theta, q) := \int_0^\pi \left|\sin\frac{\omega}{2}\right|^{(\theta - 1)q} \left|\sin\left(\frac{\omega}{2} - \phi\right)\right|^{-q\theta} d\omega. \tag{3.36}$$

Here, the integral on the interval $(\pi, 2\pi)$ was transformed to the interval $(0, \pi)$ by the substitution $\psi = 2\pi - \omega$.

The function (3.36) has the properties:

$$P(\phi + \pi, \theta, q) = P(\phi, \theta, q), \qquad P(-\phi, \theta, q) = P(\phi, 1 - \theta, q). \tag{3.37}$$

We consider the case $0 < \phi < \pi/2$ and minorize the sine function by a piecewise linear function with respect to ω:

$$\sin \frac{\omega}{2} \geq \frac{\omega}{\pi}, \qquad 0 \leq \omega \leq \pi,$$

$$\left| \sin \left(\frac{\omega}{2} - \phi \right) \right| \geq \begin{cases} \frac{2\phi - \omega}{2\phi} \sin \phi, & 0 \leq \omega \leq 2\phi, \\ \frac{\omega - 2\phi}{\pi - 2\phi} \cos \phi, & 2\phi \leq \omega \leq \pi. \end{cases}$$

Inserting this termes into (3.36), we obtain with the substitution $\omega = 2\tau\phi$:

$$P(\phi, \theta, q) \leq P_1(\phi, \theta, q) + P_2(\phi, \theta, q),$$

$$P_1(\phi, \theta, q) := \frac{(2\phi)^{1+\theta q - q}(\sin \phi)^{-\theta q}}{\pi^{(\theta-1)q}} \underbrace{\int_0^1 \tau^{(\theta-1)q}(1-\tau)^{-\theta q} d\tau}_{= B(1 + \theta q - q, 1 - \theta q)},$$

$$P_2(\phi, \theta, q) := \frac{(2\phi)^{1-q}}{\pi^{(\theta-1)q}} \left(\frac{\cos \phi}{\pi - 2\phi} \right)^{-\theta q} \int_1^{\frac{\pi}{2\phi}} \tau^{(\theta-1)q}(\tau - 1)^{-\theta q} d\tau.$$

$B(.,.)$ is Euler's beta function. For $\phi \to 0$, the terms P_1 and P_2 behave like ϕ^{1-q}; for $\phi \to \pi/2$, P_1 is bounded and P_2 behaves like $(\pi/2 - \phi)^{1-\theta q}$. The cases where ϕ does not belong to the interval $(0, \pi/2)$ can be reduced to the case considered just before by use of (3.37). Now we can derive from (3.35) the estimate

$$\int_0^{2\pi} \left| \widehat{G}_-(\lambda, \phi) \right|^2 d\phi \leq \tilde{C}_\theta \|g_-\|_{L^p}^2.$$

The constant $\tilde{C}_\theta$ is independent of $\sigma = \mathfrak{Re}\, \lambda$. An analogical estimate holds for the L^2 norm of the function (3.25). Inserting this into (3.24), we obtain the norm estimate

$$\left\| (Gf)(r, \phi); \, \overset{\circ}{W}^s_\iota(\mathbb{K}) \right\|^2 \leq C \left\| f(y); \, \overset{\circ}{W}^s_\gamma(\mathbb{R}_+) \right\|^2,$$

for

$$s - \frac{\gamma + 1}{2} = s - 1 - \frac{\iota}{2} \in (-1, 0); \quad s \in [0, 1).$$

The constant C depends only on s and the function $g(\phi)$. The theorem is proved. ∎

3.5.2 The Jump Relation

The series representations (3.25) and (3.26) are valid apriori only for $\phi \in (0, 2\pi)$. If we pass in (3.24) to the limits $\phi \nearrow 2\pi$, $\phi \searrow 0$ and subtract both values from each other, then we obtain the jump relation

$$\mathcal{M}[Gf](\lambda, 2\pi - 0) - \mathcal{M}[Gf](\lambda, +0) = S(g)\,\mathcal{M}[f](\lambda),$$

$$S(g) := 2\pi i \sum_{k=0}^{\infty} (g_{-k} - g_{k+1}) \ \psi_k(\lambda, 0).$$

It can easily be verified that $\psi_k(\lambda, 0) = 1$ for all k. Thus the expression $S(g)$ does not depend on λ, however, it makes only sense if the series converges. But this is not ensured by the assumptions (3.20), (3.21). The convergence must be assumed additionally. Sufficient for the existence of both limits (for almost all λ) is the fact that $g(\phi)$ belongs to the *Wiener algebra* $\mathcal{W}$. This is the set of all 2π-periodic functions with an absolute convergent Fourier series.

3.5.3 A Corollary to the Basic Theorem

Again, let $G(z) = r^{-1}g(\phi)$ be a homogeneous function of degree -1. It is assumed that the first derivative of the 2π-periodic function $g(\phi)$ has an absolutely convergent Fourier series:

$$g'(\phi) \in \mathcal{W} \iff \sum_{k=-\infty}^{\infty} |kg_k| < \infty. \tag{3.38}$$

Theorem 3.5.2. *Let Y be a positive real number and let the conditions (3.20) and (3.38) be fulfilled. Then the integral operator*

$$(G_Y f)(x) = \int_0^Y G(x_1 - y, x_2) f(y) dy \tag{3.39}$$

is a contionuous mapping from $H^s(0, Y)$ to $H_{loc}^{s+1/2}(\mathbb{K})$ for $0 \le s < 1/2$.

Proof. For $0 \le s < 1/2$ the space $H^s(0, Y)$ coincides with the restrictions of all functions from $\overset{\circ}{W}_0^s$ ($\mathbb{R}_+$) to the interval $(0, Y)$ [7]. It is possible to identify the function $f(y) \in H^s(0, Y)$ with $f^\sim$ (its continuation by zero) and to consider the operator G defined by (3.22) instead of G_Y. The complex variable z is replaced again by $x \in \mathbb{R}^2$.

Within this proof the letters α, β denote indices which take the values 1 and 2. Furthermore, $G_{,\beta}(x, y)$ denotes the partial derivative of $G(x, y)$ with respect to x_β and G_β denotes the operator (3.22) with the kernel $G_{,\beta}(x, y)$.

When the derivative with respect to x_β is applied to equation (3.22), the differentiation commutes with the integration for all $x \in \mathbb{K}$. The commutator

$$([\partial_\beta G - G_\beta] f)(x) := \partial_\beta (Gf)(x) - (G_\beta f)(x) \tag{3.40}$$

vanishes for $\beta = 1$. For $\beta = 2$ it represents, due to the jump relation, a mapping onto distributions with a support contained in $[0, \infty)$. The representation

$$\begin{pmatrix} G_{,1}(x) \\ G_{,2}(x) \end{pmatrix} = \frac{1}{r^3} \left[\begin{pmatrix} -x_2 \\ x_1 \end{pmatrix} g'(\phi) - \begin{pmatrix} x_1 \\ x_2 \end{pmatrix} g(\phi) \right]$$

shows that Theorem 3.5.1 can be applied to the four operators $x_\alpha G_\beta$. This gives

$$x_\alpha G_\beta \ : \quad \overset{\circ}{W}{}^s_\gamma(\mathbb{R}_+) \longmapsto \overset{\circ}{W}{}^s_{\gamma-1}(\mathbb{K}), \quad \text{for } |2s - \gamma| < 1; \quad s \in [0, 1).$$

We set $\gamma = 0$, apply Lemma 3.4.1 and obtain for $0 \le s < 1/2$:

$$x_\alpha G_\beta \ : \quad \overset{\circ}{W}{}^s_0(\mathbb{R}_+) \longmapsto \overset{\circ}{W}{}^s_{-1}(\mathbb{K}),$$

$$|x|^{-(s+1/2)} x_\alpha G_\beta \ : \quad \overset{\circ}{W}{}^s_0(\mathbb{R}_+) \longmapsto \overset{\circ}{W}{}^s_{2s}(\mathbb{K}) \subset L^2(\mathbb{K}).$$

The last relation can be written also by use of the norm (3.17). If one integrates only over the wedge $\mathbb{K}[\varepsilon, 2\pi - \varepsilon]$, then $G_{,\beta}$ can be replaced by $\partial_\beta G$ since the commutator contributes nothing:

$$\int_{\mathbb{K}[\varepsilon, 2\pi - \varepsilon]} |x|^{1-2s} |\partial_\beta(Gf)(x)|^2 \, dx < C_s \left\| f; \overset{\circ}{W}{}^s_0 \right\|^2, \qquad 0 \le s < \frac{1}{2}.$$

The constant C_s can be choosen independent of ε. Thus the last inequality remains valid in the limit case $\varepsilon = 0$.

Now we apply Lemma 3.4.2 with $t = 1/2 - s$ and multiply the function for large $|x|$ with a suitable cut-off function to obtain the assertion of the theorem.   ∎

We have shown that the operator (3.39) maps into the spaces H^s for $s < 1$. This upper bound cannot be improved even if the functions $g(\phi)$ and $f(y)$ are arbitrarily smooth. For instance, the operator with the kernel $r^{-1} \sin \phi$, applied to a C^∞ function, not-vanishing at the origin, gives a function which behaves in the vicinity of the origin like $\arctan x_1/x_2$. Such a discontinuity belongs (locally) to H^s for all $s < 1$, but not to H^1.

3.6 Integral Operators Considered as Pseudodifferential Operators

Definition 3.6.1. *For real numbers m, the symbol class S^m is defined as the set of all functions $a(x, \xi) \in C^\infty(\mathbb{R}^n \times \mathbb{R}^n)$ with following property [63, Def. 18.1.1]: For all multi-indices α, β there exists a constant $C_{\alpha\beta}$ such that*

$$\left| \frac{\partial^\alpha}{\partial \xi^\alpha} \frac{\partial^\beta}{\partial x^\beta} a(x, \xi) \right| \le C_{\alpha\beta} (1 + |\xi|)^{m-|\alpha|}, \qquad \forall x, \xi \in \mathbb{R}^n. \tag{3.41}$$

An operator of the form

$$(Au)(x) = \mathcal{F}^{-1}_{\xi \to x} a(x, \xi) \widehat{u}(\xi)$$

with a symbol $a(x, \xi) \in S^m$ is called a "classical pseudodifferential operator" of order m.

Lemma 3.6.1. [63, Theorem 18.1.13] *A classical pseudodifferential operator of order m is a continuous mapping from $H^s(\mathbb{R}^n)$ to $H^{s-m}(\mathbb{R}^n)$ for each real number s.*

Let $\Omega \subset \mathbb{R}^n$ be a bounded domain, let $\nu \geq 1 - n$ be an integer and let $g(x,y)$ belong to the class $\Psi\overline{\Omega}(\nu)$ introduced in Definition 2.9.1. Recall that $g(x,y)$ can be continued to a larger domain $\Omega_1 \supset \overline{\Omega}$. Denote by $\chi_1(x) \in C_0^\infty(\mathbb{R}^n)$ a cut-off function with following properties:

$$\chi_1(x) \equiv 1 \text{ in } \Omega \qquad \text{and} \qquad \chi_1(x) \equiv 0 \text{ in } \mathbb{R}^n \setminus \Omega_1. \tag{3.42}$$

Then

$$g_\chi(x,y) := \chi_1(x)\,\chi_1(y)\,g(x,y) \tag{3.43}$$

is a C^∞ function on $\{(x,y) \in \mathbb{R}^n \times \mathbb{R}^n \mid x \neq y\}$.

Theorem 3.6.1. *Let $\nu \geq 1 - n$, $g(x,y) \in \Psi\overline{\Omega}(\nu)$ and let g_χ be the function defined by (3.43). Then the two integral operators*

$$(G_\chi^* u)(y) := \int_{\mathbb{R}^n} g_\chi(x,y)u(x)dx; \quad (G_\chi u)(x) := \int_{\mathbb{R}^n} g_\chi(x,y)u(y)dy \tag{3.44}$$

are classical pseudodifferential operators of order $-(n+\nu)$.

Proof. We show the assertion for the adjoint operator G_χ^*. The proof for the operator G is analogous, after changing the roles of x and y.

We consider $g_\chi(x,y)$ as a function of the arguments y and $\vartheta = x - y$ and set

$$G(y,\vartheta) := g_\chi(y + \vartheta, y).$$

With its Fourier transform

$$\widehat{G}(y,\xi) := \mathcal{F}_{\vartheta \to \xi}G(y,\vartheta), \tag{3.45}$$

the operator under consideration can be written in the form

$$(G_\chi^* u)(y) = \mathcal{F}_{\xi \to y}^{-1}\widehat{G}(y,-\xi)\widehat{u}(\xi).$$

This is a pseudodifferential operator with the symbol $\widehat{G}(y,-\xi)$.

The property $\widehat{G}(y,\xi) \in C^\infty(\mathbb{R}^n \times \mathbb{R}^n)$ is obvious from the construction of $\widehat{G}$. It remains to show that the estimate (3.41) is satisfied with x replaced by y. From $g(x,y) \in \Psi\overline{\Omega}(\nu)$ it follows per definition that, to each fixed $y \in \overline{\Omega}$, there exists a radius of convergence $\varepsilon(y)$ for the pseudohomogeneous series expansion. It depends continuously on y and, since Ω is bounded, it has a lower bound $\varepsilon_0 > 0$. Using the cut-off function (1.65), we split the function (3.45) into the sum

$$\widehat{G}(y,\xi) = \chi_1(y)\,\mathcal{F}_{\vartheta \to \xi}\left[\chi\left(\frac{\vartheta}{\varepsilon_0}\right)g(y+\vartheta,y)\right] + $$

$$+ \chi_1(y)\,\mathcal{F}_{\vartheta \to \xi}\left\{\left[\chi_1(y+\vartheta) - \chi\left(\frac{\vartheta}{\varepsilon_0}\right)\right]g(y+\vartheta,y)\right\}.$$

Both summands are C^∞ functions with respect to y and have compact support. The second summand, considered as a function of ξ, is the Fourier transform of a C^∞ function with compact support and hence, it belongs to the space $S(\mathbb{R}^n)$. Lemma 1.12.1 can be applied to the first summand, which completes the proof. ∎

Remark 3.6.1. The parity condition was not used in the proof of Theorem 3.6.1. The local expansion into series with pseudohomogeneous parts is only needed for fixed y (i. e., in form of condition 3 from Definition 2.9.1) when the operator G^*_χ is considered. For the operator G_χ we need only the expansion for fixed x (i. e., in form of condition 4 from Definition 2.9.1).

The assertion of Theorem 3.6.1 can be carried over to the case $\nu \leq -n$, but then the integral operators must be interpreted for $\nu = -n$ in the Cauchy principal value sense and for $\nu < -n$ in the partie-finie sense, or alternatively, as integro-differential operators.

More detailed investigations of pseudodifferential operators as they occur within the framework of boundary element methods will be given in [69, Chap. 6]

3.7 Mapping Properties of Integral Operators with Pseudohomogeneous Kernels

Theorem 3.7.1. *Let* $\Omega \subset \mathbb{R}^n$ *be a bounded Lipschitz domain,* $\mu \in \mathbb{N}$ *and* $g(x,y) \in \Psi\overline{\Omega}(\mu - n)$. *Then the integral operator*

$$(G^\Omega u)(x) := \int_\Omega g(x,y)u(y)dy \qquad (3.46)$$

maps as follows:

$$G^\Omega : \quad \widetilde{H}^s(\Omega) \longmapsto H^{s+\mu}(\Omega), \qquad \forall s \in \mathbb{R}^1, \qquad (3.47)$$

$$G^\Omega : \quad H^s(\Omega) \longmapsto H^{s+\mu}(\Omega), \qquad \forall s \in \left(-\frac{1}{2}, \frac{1}{2}\right), \qquad (3.48)$$

$$G^\Omega : \quad H^s(\Omega) \hookrightarrow H^\sigma(\Omega), \qquad \forall s \geq \frac{1}{2}, \quad \forall \sigma < \frac{1}{2} + \mu, \qquad (3.49)$$

$$\mathrm{tr}_x G^\Omega : \quad \widetilde{H}^s(\Omega) \longmapsto H^{s+\mu-1/2}(\Gamma), \qquad \forall s > \frac{1}{2} - \mu, \qquad (3.50)$$

$$\mathrm{tr}_x G^\Omega : \quad H^s(\Omega) \longmapsto H^{s+\mu-1/2}(\Gamma), \qquad \forall s \in \left(-\frac{1}{2}, \frac{1}{2}\right), \qquad (3.51)$$

$$\mathrm{tr}_x G^\Omega : \quad H^s(\Omega) \hookrightarrow H^\sigma(\Gamma), \qquad for\ s \geq \frac{1}{2}, \quad \sigma < \mu. \qquad (3.52)$$

Proof. Suppose $u \in \tilde{H}^s(\Omega)$. This implies that the continuation by zero, $u^\sim$, belongs to the space $H^s(\mathbb{R}^n)$. The multiplication of (3.46) with the cut-off function χ_1 from (3.42) gives

$$\chi_1(x)\left(G^\Omega u\right)(x) = \int_{\mathbb{R}^2} \chi_1(x)g(x,y)\chi_1(y)u^\sim(y)\, dy = \left(G_\chi u^\sim\right)(x).$$

This function belongs to $H^{s+\mu}(\mathbb{R}^n)$, as can be seen from Theorem 3.6.1 and Lemma 3.6.1. Together with (3.14) and (3.15) we obtain the following norm estimate (where the same letter C is used for different constants):

$$\left\|(G^\Omega u)(x);\ H^{s+\mu}(\Omega)\right\| \le C\left\|\chi_1(x)(G^\Omega u)(x);\ H^{s+\mu}(\mathbb{R}^n)\right\| \le$$
$$\le C\left\|G_\chi;\ H^s \mapsto H^{s+\mu}\right\| \cdot \|u^\sim(x);\ H^s(\mathbb{R}^n)\| \le$$
$$\le C\left\|G_\chi;\ H^s \mapsto H^{s+\mu}\right\| \cdot \left\|u(x);\ \tilde{H}^s(\Omega)\right\|.$$

Thus, (3.47) is verified and since the spaces $H^s(\Omega)$ and $\tilde{H}^s(\Omega)$ coincide for $|s| < 1/2$ due to Lemma 3.3.2, property (3.48) is proved too.

To verify (3.49), we choose an $\varepsilon \in (0,1)$ with $\varepsilon < \mu - \sigma + 1/2$. From the compact embedding $H^s(\Omega) \subset\subset H^{1/2-\varepsilon}(\Omega)$ for $s \ge 1/2$ and the continuity of

$$G^\Omega :\ H^{1/2-\varepsilon}(\Omega) \longmapsto H^{\mu+1/2-\varepsilon}(\Omega)$$

follows (3.49). Indeed, either the image space coincides with $H^\sigma(\Omega)$ or it is compactly imbedded therein.

The mapping properties (3.50)–(3.52) are straightforward conclusions from (3.47)–(3.49) since

$$\mathbf{tr}_x :\ H^{\sigma+1/2}(\Omega) \longmapsto H^\sigma(\Gamma), \qquad \forall \sigma > 0,$$

per definition. ∎

Theorem 3.7.2. *Under the same assumptions as in Theorem 3.7.1, the integral operator*

$$(G^\Gamma u)(x) := \int_\Gamma g(x,y)u(y)d\Gamma_y \tag{3.53}$$

has the mapping property:

$$G^\Gamma :\ H^s(\Gamma) \longmapsto H^\sigma(\Omega), \qquad for\ -\mu < s < 0;\quad \sigma = s + \mu - \frac{1}{2}. \tag{3.54}$$

Proof. We use the idea from the proof of [23, Theorem 1 (i)] and write the operator G^Γ as superposition

$$G^\Gamma = \cdot|_\Omega \circ G^\Omega \circ \mathbf{tr}_y^*.$$

Hereby, $\cdot|_\Omega$ denotes the restriction operator and $\mathbf{tr}_y^*$ is the adjoint trace operator, which performs a continuous mapping

$$\mathbf{tr}_y^* \; : \quad H^s(\Gamma) \longmapsto H^{s-1/2}_{\text{comp}}(\mathbb{R}^n)$$

for $s < 0$. The composition with the mappings

$$G^\Omega \; : \quad H^{s-1/2}_{\text{comp}}(\mathbb{R}^n) \longmapsto H^\sigma(\mathbb{R}^n),$$

$$\cdot|_\Omega \; : \quad H^\sigma(\mathbb{R}^n) \longmapsto H^\sigma(\Omega), \qquad \text{for } \sigma > -\frac{1}{2},$$

implies the assertion of the theorem. ∎

The parity condition for the kernel $g(x,y)$ was not used within the proof of Theorem 3.7.2. The mapping property (3.54) holds as well for corresponding kernels without the parity condition. In the case $\mu = 1$ this would have the consequence that $(G^\Gamma u)(x)$ is in general not an integral anymore in the Cauchy principal value sense for $x \in \Gamma$. Then the function may have a singular behaviour for $x \to \Gamma$ even at smooth boundary points. Nevertheless, it belongs to the space H^σ for $\sigma = s + 1/2 < 1/2$.

If however, the membership to a space H^σ with $\sigma \geq 1/2$ is desired, then we need an assumption which ensures the existence of $(G^\Gamma u)(x)$ for $x \in \Gamma$ in the Cauchy principal value sense. The parity condition is sufficient for this purpose.

Definition 3.7.1. *The boundary Γ of a bounded, simple connected domain $\Omega \subset \mathbb{R}^2$ is called a "regular curvilinear polygonal" if each segment possesses a real-analytic parametrization and if the interiour angles at each corner belong to the open interval $(0, 2\pi)$.*

Theorem 3.7.3. *Suppose $\mu \in \mathbb{N}$, $n = 2$, and Γ be a regular curvilinear polygonal. Then the operator (3.53) with the kernel function $g(x,y) \in \varPsi\overline{\Omega}(\mu - 2)$ maps as follows:*

$$G^\Gamma \; : \quad H^s(\Gamma) \longmapsto H^{s+\mu-1/2}(\Omega), \quad \text{for } -\mu < s < \frac{1}{2},$$

$$G^\Gamma \; : \quad H^s(\Gamma) \hookrightarrow H^\sigma(\Omega), \qquad \text{for } s \geq \frac{1}{2}; \; \sigma < \mu.$$

Proof. For $s \in (-\mu, 0)$, the assertion follows from Theorem 3.7.2. Now we consider the case $\mu = 1$ and $0 \leq s < 1/2$. The integral over Γ is split into the sum of the integrals over the individual segments. Each segment can be mapped by a C^∞ diffeomorphism onto an interval. The essential properties of the kernel function, namely that it admits local series expansions with pseudohomogeneous terms as well as the parity condition, are preserved when the kernel is multiplied by the corresponding Taylor series of the diffeomorphism. If the argument y of the kernel is frozen in the vicinity of a corner point, Theorem 3.5.2 can be applied to the homogeneous part of degree -1. The pseudohomogeneous parts of higher degree, as well as the difference to the operator with non-frozen y, perform compact pertubations.

The case $\mu \geq 2$ can be reduced to the case just considered. Indeed, we take all derivatives of (3.53) up to the order $\mu - 1$. The corresponding derivatives of the kernel function $g(x, y)$ belong to $\Psi\overline{\Omega}(-1)$ by virtue of Lemma 1.13.2. Then, as we have already shown, all considered derivatives of $G^{\Gamma}u$ belong to $H^{s+1/2}(\Omega)$. Due to the definition of the spaces $H^s(\Omega)$, this implies $G^{\Gamma}u \in H^{s+\mu-1/2}(\Omega)$.

The assertion for $s \geq 1/2$ can be verified analogical to the proof of (3.49) by use of compact embeddings. ∎

Part II

Application to the Shell Model of Donnell-Vlasov-Type

4. The Differential Equations of the DV Model

The main part of Chapter 4 is devoted to the transformation of the model equations from the usual form based on covariant derivatives into a representation which contains only partial derivatives. Though these transformations are elementary, they are described here, since we didn't find them in the literature in this form, for the practical realization however, they are needed.

4.1 General Remarks to Shell Theory

What is a *shell*? Generally, this means a curved thin body made from solid material. In the framework considered here, the material is assumed to be elastic and the thickness d of the body is constant and very small compared to the radii of curvature. Technical examples are the bodywork of cars, cooling towers, the outer skin of airplanes and curved roof constructions.

A system of partial differential equations describing the behaviour of a shell under external forces is called a *shell model* if it contains only two independent variables. Likewise, it is possible to use the three-dimensional equations of elasticity. The replacement by two-dimensional model equations is a widely used method to reduce the complexity of the problem, also with respect to numerical methods.

There is a lot of literature on shell theory available and many different shell models where developed, due to different assumptions, hypotheses and requirements about the accuracy of the approximation. Starting with L. Euler (1771), who considered the vibration of a bell, many mathematicians, physicists and engineers dealt with the derivation of shell equations, whereby the investigations where based on mechanical principles (e. g. the elastic energy functional) as well as on practical observations and heuristical considerations.

Of special interest in this topic are verifications of some special shell models by an asymptotic analysis of the three-dimensional equations of elasticity for $d \to 0$. This problem has been investigated on since about 1989 in several papers of E. Sanchez-Palencia, P. G. Ciarlet, B. Miara, P. Destuynder and others.

This asymptotic analysis verifies the convergence of the three-dimensional solution to the solution of the two-dimensional shell equations under appropriate assumptions. An estimate of the modelling error, i. e., the distance of the two-dimensional solution from the three-dimensional one is not yet given. For the case of plates estimates of the modelling error in the energy norm are knows, see [8], [110], [111], [115], [116] and the papers cited there. Especially we mention the concept of "hierarchic models" which gives an optimal balance between the modelling error and the numerical complexity. It recommends to switch to higher order models in the vicinity of the boundary in order to have a sufficiently good approximation of the boundary layers.

4.2 Assumptions and Properties of the Model

Our object is the application of the BDIM to shell equations. Among the varity of different models we decided in favour of the shell model of Donnell-Vlasov-type (DV model). It seems to be the simplest model which fulfills the following requirements:

1. it leads to equations with variable coefficients;
2. it contains stretching and bending effects.

Pure membrane models do not fulfill the second item.

A description of the DV model can be found e. g. in [9, sect. 6.1]. The reason the model stands there under the headline "theory of shallow shells", is contributed to the already mentioned non-unique interpretation of this terminology. In the sense desribed in the Introduction, the DV model is applicable as well for non-shallow shells.

The DV model is a simplified version of *Koiter's model*. Both models are based on the following assumptions:

– The shell is in stationary equilibrium.
– The material is elastic, homogeneous and isotropic.
– The displacements are so small that the linear theory can be applied.
– The shell thickness d is so small compared to both the diameter L of the considered shell sector as well as to the minimal radius R of curvature, that all terms of magnitude d/L and d^2/R^2 can be neglected compared to unity:

$$1 + \frac{d}{L} \cong 1; \qquad 1 + \frac{d^2}{R^2} \cong 1. \qquad (4.1)$$

– The Kirchhoff-Love hypothesis can be applied.

Koiter's shell model is applicable to both extreme cases: the membrane dominated and the bending dominated shell. The DV model is derived from Koiter's model by cancellation of all terms which can be considered as very small, except for the pure bending state. In this way, the whole system becomes simpler, at least in the sense that it contains fewer terms. Certainly, the DV model is not applicable anymore to pure or almost pure bending states. Furthermore, the space of strain-free deformations has a more complicated structure (see sect. 4.10).

Koiter's model and the DV model are formally selfadjoint systems of partial differential equations. This property is sometimes called "consistency" of a shell model.

Both models belong to the class of the "first approximation of shell theory", which is characterized by a constitutive relation of the form (4.10) where the stress couples (components of the 1st moment of the stress tensor) equals to the product of the 1st moment of strain and the 0th moment of the three-dimensional, curvilinear elasticity tensor (4.8). If the product of the 0th moment of strains and the 1st moment of the elasticity tensor is added, one obtains the class of the so called "second approximation of shell theory".

Since models from this class pay attention to the product rule, the modelling error is smaller, provided that the correct formulas for the 1st moment of the elasticity tensor are used. But its calculation is non-trivial and different versions are published. A derivation of the asymptotically correct formulas can be found in [97, Appendix].

The second approximation leads in general to non-consistent shell models. For the FEM, consistent models are much more convenient. For the BDIM however, this property is irrelevant since non-symmetric matrices are generated anyway.

An application of the BDIM to other shell models than the DV model is of course possible, but not investigated here.

4.3 Geometrical Configuration

Most of the notations introduced below follow the monograph [9]. Let

$$\mathbf{r}(x_1, x_2) = \begin{pmatrix} X_1(x_1, x_2) \\ X_2(x_1, x_2) \\ X_3(x_1, x_2) \end{pmatrix} \tag{4.2}$$

be the position vector of the shell mid-surface. Its Cartesian components X_j depend on the two-dimensional parametric vector $x = (x_1, x_2) \in \Omega \subset \mathbb{R}^2$. This means, x_1 and x_2 are the curvilinear coordinates of the mid-surface. We suppose that X_j are real-analytic functions in a larger domain $\Omega_1 \supset \overline{\Omega}$ and that the covariant base vectors

$$\mathbf{a}_1 := \mathbf{r}_{,1}, \qquad \mathbf{a}_2 := \mathbf{r}_{,2} \tag{4.3}$$

are linearly independent everywhere in Ω. The notation $\cdot_{,\alpha}$ stands for the partial derivative with respect to x_α.

The scalar products

$$a_{\alpha\beta} := (\mathbf{a}_\alpha, \mathbf{a}_\beta)$$

form the covariant components of the *first fundamental tensor*. It is also called *metric tensor* since it reflects the areal metric. Inversion of the corresponding 2×2 system gives the contravariant components

$$\begin{pmatrix} a^{11} & a^{12} \\ a^{21} & a^{22} \end{pmatrix} = \frac{1}{a} \begin{pmatrix} a_{22} & -a_{12} \\ -a_{21} & a_{11} \end{pmatrix}.$$

For the determinant there holds Lagrange's identity

$$a := a_{11}a_{22} - a_{12}^2 = |\mathbf{a}_1 \times \mathbf{a}_2|^2.$$

The contravariant base vectors $\mathbf{a}^\alpha := a^{\alpha\beta}\mathbf{a}_\beta$ fulfill the bi-orthogonality relation $(\mathbf{a}^\alpha, \mathbf{a}_\beta) = \delta^\alpha_\beta$. Together with the unit normal vector

$$\mathbf{a}^3 = \mathbf{a}_3 := \frac{\mathbf{a}_1 \times \mathbf{a}_2}{\sqrt{a}},$$

one gets the three-dimensional covariant base $\mathbf{a}_1, \mathbf{a}_2, \mathbf{a}_3$ and the contravariant base $\mathbf{a}^1, \mathbf{a}^2, \mathbf{a}^3$, which are bi-orthogonal to each other as well.

The second fundamental tensor with the covariant components

$$b_{\alpha\beta} := \mathbf{r}_{,\alpha\beta} \cdot \mathbf{a}_3$$

contains the information on the curvature of the mid-surface.

The shell equations contain several constants, namely Young's modulus E, Poisson's ratio ν and stretching and bending stiffnesses

$$D := \frac{E\,d}{1 - \nu^2}, \qquad B := \frac{E\,d^3}{12(1 - \nu^2)}.$$

Furthermore, covariant derivatives occur, which are defined for tensor components of first and second order as follows:

$$\left. \begin{aligned} w_{\alpha|\beta} &= w_{\alpha,\beta} - \Gamma^\gamma_{\alpha\beta} w_\gamma, & w_{\alpha\beta}|_\gamma &= w_{\alpha\beta,\gamma} - \Gamma^\lambda_{\alpha\gamma} w_{\lambda\beta} - \Gamma^\lambda_{\beta\gamma} w_{\alpha\lambda}, \\ w^\alpha|_\beta &= w^\alpha{}_{,\beta} + \Gamma^\alpha_{\gamma\beta} w^\gamma, & w^{\alpha\beta}|_\gamma &= w^{\alpha\beta}{}_{,\gamma} + \Gamma^\alpha_{\gamma\lambda} w^{\lambda\beta} + \Gamma^\beta_{\gamma\lambda} w^{\alpha\lambda}. \end{aligned} \right\} \quad (4.4)$$

The Christoffel symbols $\Gamma^\gamma_{\alpha\beta} := \mathbf{a}_{\alpha,\beta} \cdot \mathbf{a}^\gamma$ fulfill the property

$$\Gamma^\alpha_{\alpha\beta} = \frac{1}{\sqrt{a}} \left(\sqrt{a} \right)_{,\beta}. \qquad (4.5)$$

4.4 Shell Equations in Covariant Derivatives

Let

$$\mathbf{v}(x_1, x_2) = v_j(x_1, x_2) \cdot \mathbf{a}^j(x_1, x_2)$$

denote the displacement vector of the mid-surface. Its covariant components v_j are related to strain components via

$$\alpha_{(\alpha\beta)} = \frac{1}{2} \left(v_{\alpha|\beta} + v_{\beta|\alpha} \right) - b_{\alpha\beta} v_3, \qquad (4.6)$$

$$\omega_{(\alpha\beta)} = -v_{3|\alpha\beta}. \qquad (4.7)$$

Multiplication with the isotropic elasticity tensor

$$H^{\alpha\beta\lambda\mu} := \frac{1-\nu}{2} \left(a^{\alpha\lambda}a^{\beta\mu} + a^{\alpha\mu}a^{\beta\lambda} + \frac{2\nu}{1-\nu}a^{\alpha\beta}a^{\lambda\mu} \right) \qquad (4.8)$$

gives the symmetric stress resultants

$$n^{(\alpha\beta)} = D\,H^{\alpha\beta\lambda\mu}\,\alpha_{(\lambda\mu)} \qquad (4.9)$$

and the symmetric stress couples

$$m^{(\alpha\beta)} = B\,H^{\alpha\beta\lambda\mu}\,\omega_{(\lambda\mu)}. \qquad (4.10)$$

The equilibrium relations

$$n^{(\alpha\beta)}\Big|_{\beta} = -p^{\alpha} \qquad (4.11)$$

$$q^{\alpha}|_{\alpha} + b_{\alpha\beta}\,n^{(\alpha\beta)} = -p^{3}. \qquad (4.12)$$

complete the model. Here

$$q^{\alpha} := m^{(\alpha\beta)}|_{\beta} \qquad (4.13)$$

denote the shear forces and p^{j} are the contravariant components of the surface load

$$\mathbf{p}(x_1, x_2) = p^{j}(x_1, x_2) \cdot \mathbf{a}_{j}(x_1, x_2).$$

The above form is the usual one in shell theory. For our purposes however, it is more convenient to multiply equations (4.11) and (4.12) by $\sqrt{a}$. Inserting the equations (4.6)–(4.13) into each other gives a 3×3 system of linear partial differential equations:

$$\begin{pmatrix} A^{11} & A^{12} & A^{13} \\ A^{21} & A^{22} & A^{23} \\ A^{31} & A^{32} & A^{33} \end{pmatrix} \begin{pmatrix} v_1 \\ v_2 \\ v_3 \end{pmatrix} = -\sqrt{a} \begin{pmatrix} p^1 \\ p^2 \\ p^3 \end{pmatrix}. \qquad (4.14)$$

If we replace the covariant derivatives in the above equations according to (4.4) by partial derivatives and Christoffel symbols, then we obtain after multiplying out very long expressions which contain derivatives of the Christoffel symbols. The next sections are devoted to the derivation of compact formulas for the operators A^{jk}. This form is more convenient for the practical construction of Levi functions and for the calculation of the associated integral kernels. It appears that all Christoffel symbols can be eliminated from the equations. (This observation is not valid for Koiter's model.)

4.5 Preliminaries from Differential Geometry

The operations of raising and lowering of indices are performed by multiplication with the first fundamental tensor. For instance,

$$b^\alpha_\beta := a^{\alpha\lambda} b_{\lambda\beta}$$

are the mixed components of the second fundamental tensor. If an index appears twice (which means summation), the lower index can be raised and the upper index can be lowered at the same time. This basic procedure is sometimes used in what follows.

First we introduce some further notations: the permutation tensor

$$\begin{pmatrix} \varepsilon^{11} & \varepsilon^{12} \\ \varepsilon^{21} & \varepsilon^{22} \end{pmatrix} := \frac{1}{\sqrt{a}} \begin{pmatrix} 0 & 1 \\ -1 & 0 \end{pmatrix}, \qquad \varepsilon_{\alpha\beta} := a\varepsilon^{\alpha\beta}, \qquad (4.15)$$

the mean curvature

$$\mathcal{H} := \frac{1}{2} b^\lambda_\lambda = \frac{1}{2} a^{\lambda\mu} b_{\lambda\mu} \qquad (4.16)$$

and the Gaussian curvature

$$\mathcal{K} := b^1_1 b^2_2 - b^1_2 b^2_1 = \frac{1}{2} \varepsilon^{\alpha\lambda} \varepsilon^{\beta\mu} b_{\lambda\mu} b_{\alpha\beta}. \qquad (4.17)$$

Further we need the following 2×2 matrices

$$\underline{M} := \sqrt{a} \begin{pmatrix} a^{11} & a^{12} \\ a^{21} & a^{22} \end{pmatrix}, \qquad (4.18)$$

$$\underline{N} := \begin{pmatrix} b^2_2 & -b^2_1 \\ -b^1_2 & b^1_1 \end{pmatrix}. \qquad (4.19)$$

The matrix $\underline{M}$ is symmetric and its determinant is 1. The product $\underline{M} \cdot \underline{N}$ is also symmetric but the matrix $\underline{N}$ is non-symmetric in general. Its entries can be written in the form

$$N^{\cdot\beta}_{\gamma\cdot} = \varepsilon_{\gamma\lambda} \varepsilon^{\beta\mu} b^\lambda_\mu \qquad (4.20)$$

where γ is the row index and β is the column index.

The replacement of covariant derivatives by partial ones is based on the relation

$$w^\alpha |_\alpha = \frac{1}{\sqrt{a}} \left(\sqrt{a}\, w^\alpha \right)_{,\alpha} \qquad (4.21)$$

which is obvious from (4.4) and (4.5). Further we need Ricci's lemma sometimes, which says that covariant derivatives of the metric tensor and the permutation tensor vanish. Henceforce, they can be handled like constant factors if the product rule for covariant derivatives is applied.

Ricci's lemma is used in the first step of the following calculation. For the components w_λ of an arbitrary vector we receive:

$$\left(\varepsilon^{\beta\mu}w_{\lambda|\mu}\right)|_{\beta} = \varepsilon^{\beta\mu}w_{\lambda|\mu\beta} = \frac{1}{\sqrt{a}}\left(w_{\lambda|21} - w_{\lambda|12}\right)$$

$$= \frac{1}{\sqrt{a}}R^{\rho}_{.\lambda21}w_{\rho} = \frac{1}{\sqrt{a}}R_{\rho\lambda21}w^{\rho}.$$

Here, $R^{\rho}_{.\lambda\xi\eta}$ denotes the Riemannian curvature tensor which can be defined by various ways, e. g., by the commutator relation

$$w_{\lambda|\xi\eta} - w_{\lambda|\eta\xi} = R^{\rho}_{.\lambda\xi\eta}w_{\rho}.$$

Due to the *theorema egregium* of Gauss there holds:

$$R_{\rho\lambda12} = a_{\rho\eta}R^{\eta}_{.\lambda12} = \left\{\begin{array}{lll} 0 & \text{, für} & \rho = \lambda \\ a\,\mathcal{K} & \text{, für} & \rho = 1, \quad \lambda = 2 \\ -a\,\mathcal{K} & \text{, für} & \rho = 2, \quad \lambda = 1 \end{array}\right.$$

This gives the formula

$$\varepsilon^{\beta\mu}w_{\lambda|\mu\beta} = \mathcal{K}\,\varepsilon_{\lambda\rho}w^{\rho}. \tag{4.22}$$

Further we need in the sequel the equations of Mainardi and Codazzi:

$$\varepsilon^{\beta\mu}b_{\lambda\mu|\beta} = 0 \tag{4.23}$$

and the relation

$$\varepsilon^{\beta\mu}w_{\mu|\beta} = \varepsilon^{\beta\mu}w_{\mu,\beta} \tag{4.24}$$

which is obvious from the definition (4.4) of the covariant derivative.

4.6 Splitting of the Shell Equations

The permutation tensor (4.15) fulfills the relation

$$\varepsilon^{\alpha\lambda}\varepsilon^{\beta\mu} + \varepsilon^{\alpha\mu}\varepsilon^{\beta\lambda} = 2\,a^{\alpha\beta}a^{\lambda\mu} - a^{\alpha\mu}a^{\beta\lambda} - a^{\alpha\lambda}a^{\beta\mu}$$

(cmp. [9, formula 1.3.12]). This suggests the splitting

$$H^{\alpha\beta\lambda\mu} = \hat{H}^{\alpha\beta\lambda\mu} + (1 - \nu)\tilde{H}^{\alpha\beta\lambda\mu} \tag{4.25}$$

of the elasticity tensor (4.8), whereby

$$\hat{H}^{\alpha\beta\lambda\mu} := a^{\alpha\beta}a^{\lambda\mu},$$
$$\tilde{H}^{\alpha\beta\lambda\mu} := -\frac{1}{2}\left(\varepsilon^{\alpha\lambda}\varepsilon^{\beta\mu} + \varepsilon^{\alpha\mu}\varepsilon^{\beta\lambda}\right). \tag{4.26}$$

According to (4.25), the operators A^{jk}, the shear forces q^{α} and the tensors $n^{(\alpha\beta)}$, $m^{(\alpha\beta)}$ are split into:

$$A^{jk} = \hat{A}^{jk} + (1 - \nu)\tilde{A}^{jk}, \tag{4.27}$$
$$n^{(\alpha\beta)} = \hat{n}^{(\alpha\beta)} + (1 - \nu)\tilde{n}^{(\alpha\beta)},$$
$$m^{(\alpha\beta)} = \hat{m}^{(\alpha\beta)} + (1 - \nu)\tilde{m}^{(\alpha\beta)},$$
$$q^{\alpha} = \hat{q}^{\alpha} + (1 - \nu)\tilde{q}^{\alpha}.$$

Both parts of the operators (4.27) at first are treated separately, afterwards, in section 4.7 they are added.

The splitting of the stresses and moments has the following physical interpretation: the terms with a hat correspond to uniform compressions or de-compressions, while the tilde denotes deviatoric terms.

4.6.1 The First Part

From (4.16), (4.21) and Ricci's lemma we obtain

$$
\begin{aligned}
\hat{n}^{(\alpha\beta)} &= D\, a^{\alpha\beta} a^{\lambda\mu} \alpha_{(\lambda\mu)} = D\, a^{\alpha\beta} a^{\lambda\mu} \left[\frac{1}{2}\left(v_{\lambda|\mu} + v_{\mu|\lambda}\right) - b_{\lambda\mu} v_3\right] \\
&= D\, a^{\alpha\beta} \left[\frac{1}{2}\left(a^{\lambda\mu} v_\lambda\right)|_\mu + \frac{1}{2}\left(a^{\lambda\mu} v_\mu\right)|_\lambda - 2\mathcal{H} v_3\right] \\
&= D\, a^{\alpha\beta} \left(v^\mu|_\mu - 2\mathcal{H} v_3\right) \\
&= D\, a^{\alpha\beta} \left[\frac{1}{\sqrt{a}}\left(\sqrt{a}\, v^\mu\right)_{,\mu} - 2\mathcal{H}\, v_3\right]
\end{aligned}
\tag{4.28}
$$

and

$$
\begin{aligned}
\hat{m}^{(\alpha\beta)} &= -Ba^{\alpha\beta} a^{\lambda\mu} v_{3|\lambda\mu} = -Ba^{\alpha\beta}\left(a^{\lambda\mu} v_{3|\lambda}\right)|_\mu \\
&= -Ba^{\alpha\beta}\left(a^{\lambda\mu} v_{3,\lambda}\right)|_\mu \\
&= -Ba^{\alpha\beta} \frac{1}{\sqrt{a}}\left(\sqrt{a}\, a^{\lambda\mu} v_{3,\lambda}\right)_{,\mu}.
\end{aligned}
\tag{4.29}
$$

The first term on the left-hand side of the equilibrium relation (4.11) takes the form

$$
\hat{n}^{(\alpha\beta)}|_\beta = D\, a^{\alpha\beta} \left[\frac{1}{\sqrt{a}}\left(\sqrt{a}\, v^\mu\right)_{,\mu} - 2\mathcal{H} v_3\right]_{,\beta}.
\tag{4.30}
$$

Hereby we have again used Ricci's lemma and the fact that the covariant derivative of a scalar function coincides with the partial derivative. The same argument leads to

$$
\hat{q}^{\alpha} = \hat{m}^{\alpha\beta}|_\beta = -B\, a^{\alpha\beta} \left[\frac{1}{\sqrt{a}}\left(\sqrt{a}\, a^{\lambda\mu} v_{3,\lambda}\right)_{,\mu}\right]_{,\beta}.
$$

Finally, from (4.16) and

$$\hat{q}^{\alpha}|_{\alpha} = \frac{1}{\sqrt{a}}\left(\sqrt{a}\,\hat{q}^{\alpha}\right)_{,\alpha}, \tag{4.31}$$

$$b_{\alpha\beta}\,\hat{n}^{(\alpha\beta)} = \frac{2D\mathcal{H}}{\sqrt{a}}\left(\sqrt{a}\,v^{\mu}\right)_{,\mu} - 4\,D\,\mathcal{H}^{2}\,v_{3} \tag{4.32}$$

(see (4.21) and (4.16)) we obtain the following matrix representations:

$$\begin{pmatrix} \hat{A}^{11} & \hat{A}^{12} \\ \hat{A}^{21} & \hat{A}^{22} \end{pmatrix} = D\,\underline{M}\begin{pmatrix} \partial_{1} \\ \partial_{2} \end{pmatrix}\frac{1}{\sqrt{a}}\begin{pmatrix} \partial_{1} \\ \partial_{2} \end{pmatrix}^{\mathsf{T}}\underline{M}, \tag{4.33}$$

$$\begin{pmatrix} \hat{A}^{13} \\ \hat{A}^{23} \end{pmatrix} = -2D\,\underline{M}\begin{pmatrix} \partial_{1} \\ \partial_{2} \end{pmatrix}\mathcal{H}, \tag{4.34}$$

$$\begin{pmatrix} \hat{A}^{31} \\ \hat{A}^{32} \end{pmatrix}^{\mathsf{T}} = 2D\mathcal{H}\begin{pmatrix} \partial_{1} \\ \partial_{2} \end{pmatrix}^{\mathsf{T}}\underline{M}, \tag{4.35}$$

$$\hat{A}^{33} = -B\Delta_{M}\frac{1}{\sqrt{a}}\Delta_{M} - 4\sqrt{a}D\mathcal{H}^{2}. \tag{4.36}$$

The notation Δ_{M} stands for the generalized Laplacian

$$\Delta_{M} := \partial_{\alpha}\left(\sqrt{a}\,a^{\alpha\beta}\partial_{\beta}\right) = \begin{pmatrix} \partial_{1} \\ \partial_{2} \end{pmatrix}^{\mathsf{T}}\underline{M}\begin{pmatrix} \partial_{1} \\ \partial_{2} \end{pmatrix}. \tag{4.37}$$

4.6.2 The Second Part

Now the part (4.26) of the elasticity tensor is inserted into the constitutive relations (4.9) and (4.10). Due to the symmetry of the tensors $\alpha_{(\alpha\beta)}$ and $\omega_{(\alpha\beta)}$, the following simplifications are possible:

$$\tilde{n}^{(\alpha\beta)} = D\,\tilde{H}^{\alpha\beta\lambda\mu}\alpha_{(\lambda\mu)} = -D\,\varepsilon^{\alpha\lambda}\varepsilon^{\beta\mu}\alpha_{(\lambda\mu)}, \tag{4.38}$$

$$\tilde{m}^{(\alpha\beta)} = B\,\tilde{H}^{\alpha\beta\lambda\mu}\omega_{(\lambda\mu)} = -B\,\varepsilon^{\alpha\lambda}\varepsilon^{\beta\mu}\omega_{(\mu\lambda)} = B\,\varepsilon^{\alpha\lambda}\varepsilon^{\beta\mu}v_{3|\mu\lambda}. \tag{4.39}$$

From the identity

$$\varepsilon^{\beta\mu}v_{\mu|\lambda\beta} = \varepsilon^{\beta\mu}\left(v_{\lambda|\mu\beta} + v_{\mu|\beta\lambda}\right), \tag{4.40}$$

which can be verified by pure index calculations, together with Ricci's lemma and the product rule for covariant derivatives, we conclude:

$$\begin{aligned} \tilde{n}^{(\alpha\beta)}|_{\beta} &= -D\,\varepsilon^{\alpha\lambda}\varepsilon^{\beta\mu}\left[\frac{1}{2}\left(v_{\lambda|\mu} + v_{\mu|\lambda}\right) - b_{\lambda\mu}v_{3}\right]_{|\beta} \\ &= -D\varepsilon^{\alpha\lambda}\varepsilon^{\beta\mu}\left[\frac{1}{2}\left(v_{\lambda|\mu\beta} + v_{\mu|\lambda\beta}\right) - \left(b_{\lambda\mu}v_{3}\right)_{|\beta}\right] \\ &= -D\varepsilon^{\alpha\lambda}\varepsilon^{\beta\mu}\left(v_{\lambda|\mu\beta} + \frac{1}{2}v_{\mu|\beta\lambda} - b_{\lambda\mu|\beta}v_{3} - b_{\lambda\mu}v_{3,\beta}\right). \end{aligned}$$

By virtue of (4.22), (4.23) and (4.24), this can be written as

$$\tilde{n}^{(\alpha\beta)}|_\beta = -D\,\mathcal{K}\,\varepsilon^{\alpha\lambda}\varepsilon_{\lambda\rho}v^\rho - \frac{D}{2}\varepsilon^{\alpha\lambda}\left(\varepsilon^{\beta\mu}v_{\mu,\beta}\right)_{,\lambda} + D\,\varepsilon^{\alpha\lambda}\varepsilon^{\beta\mu}b_{\lambda\mu}v_{3,\beta}.$$

Since

$$\varepsilon^{\alpha\lambda}\varepsilon_{\lambda\rho} = -\delta^\alpha_\rho, \tag{4.41}$$

we can write the upper left block of the matrix operator in the form

$$\begin{pmatrix} \tilde{A}^{11} & \tilde{A}^{12} \\ \tilde{A}^{21} & \tilde{A}^{22} \end{pmatrix} = D\mathcal{K}\,\underline{M} + \frac{D}{2}\begin{pmatrix} -\partial_2 \\ \partial_1 \end{pmatrix}\frac{1}{\sqrt{a}}\begin{pmatrix} -\partial_2 \\ \partial_1 \end{pmatrix}^{\mathsf{T}}. \tag{4.42}$$

The upper right block is

$$\begin{pmatrix} \tilde{A}^{13} \\ \tilde{A}^{23} \end{pmatrix} = D\,\underline{M}\,\underline{N}\begin{pmatrix} \partial_1 \\ \partial_2 \end{pmatrix}, \tag{4.43}$$

which can be deduced from

$$\varepsilon^{\alpha\lambda}\varepsilon^{\beta\mu}b_{\lambda\mu} = \varepsilon^\alpha_\lambda\varepsilon^{\beta\mu}b^\lambda_\mu = a^{\alpha\gamma}\varepsilon_{\gamma\lambda}\varepsilon^{\beta\mu}b^\lambda_\mu = a^{\alpha\gamma}N^{\cdot\beta}_\gamma.$$

(compare (4.20)).

Inserting (4.38) and (4.39) into the left-hand side of the equilibrium relations (4.12) gives

$$\tilde{m}^{(\alpha\beta)}|_{\beta\alpha} + b_{\alpha\beta}\tilde{n}^{(\alpha\beta)} = \Sigma_1 + \Sigma_2 + \Sigma_3. \tag{4.44}$$

The three summands

$$\begin{aligned}
\Sigma_1 &= B\left(\varepsilon^{\alpha\lambda}\varepsilon^{\beta\mu}v_{3|\mu\lambda}\right)|_{\beta\alpha} \\
\Sigma_2 &= -\frac{D}{2}\varepsilon^{\alpha\lambda}\varepsilon^{\beta\mu}b_{\beta\alpha}\left(v_{\lambda|\mu} + v_{\mu|\lambda}\right) \\
\Sigma_3 &= D\,\varepsilon^{\alpha\lambda}\varepsilon^{\beta\mu}b_{\lambda\mu}b_{\beta\alpha}v_3
\end{aligned}$$

are calculated separately in the sequel.

For the first summand we need the identity (4.40), with v_μ replaced by $v_{3|\mu}$, and the relation

$$\varepsilon^{\beta\mu}v_{3|\mu\beta} = \varepsilon^{\beta\mu}v_{3,\mu\beta} = 0 \tag{4.45}$$

deduced from (4.24). Furthermore, (4.21), (4.22) and Ricci's lemma are used to obtain

$$\begin{aligned}
\Sigma_1 &= \frac{B}{\sqrt{a}}\left(\sqrt{a}\,\varepsilon^{\alpha\lambda}\varepsilon^{\beta\mu}v_{3|\mu\lambda\beta}\right)_{,\alpha} = \frac{B}{\sqrt{a}}\left[\sqrt{a}\,\varepsilon^{\alpha\lambda}\varepsilon^{\beta\mu}\left(v_{3|\lambda\mu\beta} + v_{3|\mu\beta\lambda}\right)\right]_{,\alpha} \\
&= \frac{B}{\sqrt{a}}\left[\sqrt{a}\,\varepsilon^{\alpha\lambda}\,\mathcal{K}\varepsilon_{\lambda\rho}a^{\rho\gamma}v_{3|\gamma} + \sqrt{a}\,\varepsilon^{\alpha\lambda}\left(\varepsilon^{\beta\mu}v_{3|\mu\beta}\right)_{|\lambda}\right]_{,\alpha} \\
&= \frac{B}{\sqrt{a}}\left(\sqrt{a}\,\varepsilon^{\alpha\lambda}\varepsilon_{\lambda\rho}\,\mathcal{K}\,a^{\rho\gamma}v_{3,\gamma}\right)_{,\alpha}.
\end{aligned}$$

With (4.41), this gives

$$\Sigma_1 = -\frac{B}{\sqrt{a}} \left(K \sqrt{a} \, a^{\alpha\gamma} v_{3,\gamma} \right)_{,\alpha} .$$

Due to (4.17), the third summand is

$$\Sigma_3 = 2 \, D \, K \, v_3 .$$

Thus we can write

$$\tilde{A}^{33} = -B \left(\begin{array}{c} \partial_1 \\ \partial_2 \end{array} \right)^{\mathsf{T}} K \, \underline{M} \left(\begin{array}{c} \partial_1 \\ \partial_2 \end{array} \right) + 2\sqrt{a} \, D \, K . \tag{4.46}$$

Since $b_{\alpha\beta}$ is symmetric, the second summand is

$$\Sigma_2 = -D \, \varepsilon^{\alpha\lambda} \varepsilon^{\beta\mu} b_{\alpha\beta} v_{\lambda|\mu} .$$

Now we apply the chain rule for covariant derivatives, Ricci's lemma and the equation (4.23) of Mainardi and Codazzi, to obtain:

$$\begin{aligned}
\Sigma_2 &= -D \, \varepsilon^{\alpha\lambda} \varepsilon^{\beta\mu} \left[(b_{\alpha\beta} v_\lambda) |_\mu - b_{\alpha\beta|\mu} v_\lambda \right] \\
&= -D \, \varepsilon^{\alpha\lambda} \varepsilon^{\beta\mu} (b_{\alpha\beta} v_\lambda) |_\mu = -D \, \varepsilon^{\beta\mu} \left(\varepsilon^{\alpha\lambda} b_{\alpha\beta} v_\lambda \right) |_\mu \\
&= -D \, \varepsilon^{\beta\mu} \left(\varepsilon^{\alpha\lambda} b_{\alpha\beta} v_\lambda \right)_{,\mu} .
\end{aligned}$$

In the last step, the covariant derivative was transformed into a partial one according to (4.24). Since the terms $\sqrt{a} \, \varepsilon^{\beta\mu}$ are constant, they commute with the differentiation. By use of (4.20) we can write

$$\begin{aligned}
\Sigma_2 &= -\frac{D}{\sqrt{a}} \left(\sqrt{a} \, \varepsilon^{\beta\mu} \varepsilon^{\alpha\lambda} b_{\alpha\beta} v_\lambda \right)_{,\mu} = -\frac{D}{\sqrt{a}} \left(\sqrt{a} \, \varepsilon^{\mu\beta} \varepsilon^{\lambda\alpha} b_{\alpha\beta} v_\lambda \right)_{,\mu} \\
&= -\frac{D}{\sqrt{a}} \left(\sqrt{a} \, \varepsilon^{\mu\beta} \varepsilon_{\lambda\alpha} b^\alpha_\beta v^\lambda \right)_{,\mu} = -\frac{D}{\sqrt{a}} \left(\sqrt{a} \, N^{\cdot\mu}_{\lambda\cdot} v^\lambda \right)_{,\mu}
\end{aligned}$$

to obtain

$$\left(\begin{array}{c} \tilde{A}^{31} \\ \tilde{A}^{32} \end{array} \right)^{\mathsf{T}} = -D \left(\begin{array}{c} \partial_1 \\ \partial_2 \end{array} \right)^{\mathsf{T}} \underline{N}^{\mathsf{T}} \, \underline{M} . \tag{4.47}$$

All parts of the matrix operator (4.14) are transformed now into the desired form. Collecting the expressions (4.33) to (4.36), (4.42), (4.43), (4.46), (4.47) and insertion into (4.27) gives the desired form of the shell equations.

4.7 The Shell Equations in Partial Derivatives

The shell model of Donnell-Vlasov-type can be written as a 3×3 system of linear partial differential equations of the form

$$\begin{pmatrix} A^{11} & A^{12} & A^{13} \\ A^{21} & A^{22} & A^{23} \\ A^{31} & A^{32} & A^{33} \end{pmatrix} \begin{pmatrix} v_1 \\ v_2 \\ v_3 \end{pmatrix} = -\sqrt{a} \begin{pmatrix} p^1 \\ p^2 \\ p^3 \end{pmatrix}. \tag{4.48}$$

Recall that v_j are the covariant components of the displacement vector and p^j are the contravariant components of the surface load. The individual operators A^{jk} have the representations:

$$A^{\square} := \begin{pmatrix} A^{11} & A^{12} \\ A^{21} & A^{22} \end{pmatrix} = \tag{4.49}$$

$$= D \left\{ \underline{M} \begin{pmatrix} \partial_1 \\ \partial_2 \end{pmatrix} \frac{1}{\sqrt{a}} \begin{pmatrix} \partial_1 \\ \partial_2 \end{pmatrix}^{\mathsf{T}} \underline{M} + \tilde{\nu} \left[\begin{pmatrix} -\partial_2 \\ \partial_1 \end{pmatrix} \frac{1}{\sqrt{a}} \begin{pmatrix} -\partial_2 \\ \partial_1 \end{pmatrix}^{\mathsf{T}} + 2\mathcal{K}\underline{M} \right] \right\}$$

$$\begin{pmatrix} A^{13} \\ A^{23} \end{pmatrix} = 2D\,\underline{M} \left[-\begin{pmatrix} \partial_1 \\ \partial_2 \end{pmatrix} \mathcal{H} + \tilde{\nu}\,\underline{N} \begin{pmatrix} \partial_1 \\ \partial_2 \end{pmatrix} \right] \tag{4.50}$$

$$\begin{pmatrix} A^{31} \\ A^{32} \end{pmatrix}^{\mathsf{T}} = 2D \left[\mathcal{H} \begin{pmatrix} \partial_1 \\ \partial_2 \end{pmatrix}^{\mathsf{T}} - \tilde{\nu} \begin{pmatrix} \partial_1 \\ \partial_2 \end{pmatrix}^{\mathsf{T}} \underline{N}^{\mathsf{T}} \right] \underline{M} \tag{4.51}$$

$$A^{33} = -B \left[\Delta_M \frac{1}{\sqrt{a}} \Delta_M + 2\tilde{\nu} \begin{pmatrix} \partial_1 \\ \partial_2 \end{pmatrix}^{\mathsf{T}} \mathcal{K}\,\underline{M} \begin{pmatrix} \partial_1 \\ \partial_2 \end{pmatrix} \right] + 4D\sqrt{a}\left(\tilde{\nu}\mathcal{K} - \mathcal{H}^2 \right) \tag{4.52}$$

The operator Δ_M is defined by (4.37). We recapitulate the meaning of the other notations. The matrices

$$\underline{M} = \sqrt{a} \begin{pmatrix} a^{11} & a^{12} \\ a^{21} & a^{22} \end{pmatrix}, \qquad \underline{N} = \begin{pmatrix} b_2^2 & -b_1^2 \\ -b_2^1 & b_1^1 \end{pmatrix}.$$

and the functions $\mathcal{H}$ (= mean curvature) and $\mathcal{K}$ (= Gaussian curvature) contain the geometric information on the mid-surface. Material properties and the shell thickness enter into the constants D (= streching stiffness), B (= bending stiffness) and

$$\tilde{\nu} := \frac{1 - \nu}{2}. \tag{4.53}$$

4.8 Shell Equations in Isothermal Parametrization

An isothermal parametrization is defined by:

$$a^{12}(x_1, x_2) \;\equiv\; 0, \tag{4.54}$$
$$a^{11}(x_1, x_2) \;\equiv\; a^{22}(x_1, x_2). \tag{4.55}$$

In this case the Christoffel symbols have the special form

$$\Gamma_{11}^1 = \Gamma_{12}^2 = -\Gamma_{22}^1 = \Lambda_{,1}, \qquad \Gamma_{22}^2 = \Gamma_{12}^1 = -\Gamma_{11}^2 = \Lambda_{,2} \tag{4.56}$$

with $\Lambda := \frac{1}{4} \ln a$.

Each sufficiently smooth curved surface possesses an isothermal parametrization. This result goes back to C. F. Gauss. It is necessary to solve the corresponding Beltrami's equation. But a closed formula for the solution can only be obtained in special cases. In general one can only deduce from the parametric representation (4.2) an approximation of the isothermal parametrization, e. g., in form of Taylor series.

With an isothermal parametrization, the system (4.48) takes a much simpler form. The matrix $\underline{M}$ becomes then the identity and the upper left block (4.49) reduces to:

$$A^\square = D \left[\begin{pmatrix} \partial_1 \\ \partial_2 \end{pmatrix} \frac{1}{\sqrt{a}} \begin{pmatrix} \partial_1 \\ \partial_2 \end{pmatrix}^{\mathsf{T}} + \tilde\nu \begin{pmatrix} -\partial_2 \\ \partial_1 \end{pmatrix} \frac{1}{\sqrt{a}} \begin{pmatrix} -\partial_2 \\ \partial_1 \end{pmatrix}^{\mathsf{T}} + 2\tilde\nu \mathcal{K}\, I \right]$$

$$= D \left[\begin{pmatrix} \partial_1 & -\partial_2 \\ \partial_2 & \partial_1 \end{pmatrix} \frac{1}{\sqrt{a}} \begin{pmatrix} 1 & 0 \\ 0 & \tilde\nu \end{pmatrix} \begin{pmatrix} \partial_1 & \partial_2 \\ -\partial_2 & \partial_1 \end{pmatrix} + 2\tilde\nu \mathcal{K}\, I \right]. \tag{4.57}$$

The other operators simplify as well if isothermal parameters are used. The operator (4.52) becomes

$$A^{33} = -B \left[\Delta \frac{1}{\sqrt{a}} \Delta + 2\tilde\nu \left(\partial_1 \mathcal{K} \partial_1 + \partial_2 \mathcal{K} \partial_2 \right) \right] + 4\sqrt{a}\, D \left(\tilde\nu \mathcal{K} - \mathcal{H}^2 \right). \tag{4.58}$$

4.9 Factorization with Respect to Cauchy-Riemann Operators

(This section is only of theoretical interest and will not be needed later.)

Formula (4.57) suggests the question, whether or not other parametrizations can lead to a similar factorization of the principal part of $A^\square$. If this would be possible, it could be employed for analytical investigations and for numerical purposes.

Definition 4.9.1. *Let A be a 2×2 matrix of partial differential operators of order 2 in two independent variables. We say, the matrix A admits a factorization with respect to Cauchy-Riemann operators (CR factorization), if it can be represented in the form*

$$A = M_1 \begin{pmatrix} \partial_1 & -\partial_2 \\ \partial_2 & \partial_1 \end{pmatrix} M_2 \begin{pmatrix} \partial_1 & \partial_2 \\ -\partial_2 & \partial_1 \end{pmatrix} M_3 + \mathcal{O}\left(\partial^1\right) \tag{4.59}$$

with certain multiplication operators M_1, M_2, M_3, i. e., 2×2 matrices with constants or functions as entries. The symbol $\mathcal{O}(\partial^1)$ stands for differential operators of order 1 and 0.

In the previous section we have shown that the upper left block (4.49) admits a CR factorization if the parametrization is isothermal. Obviously, a CR factorization would be also valid in case $\nu = 1$ for an arbitrary parametrization. This however, is not possible for physical reasons. Poisson's ratio ν is always smaller than $1/2$.

The question formulated above gets a negative answer by the following two lemmata, which are given here without proof.

Lemma 4.9.1. *Assume that $\nu \neq 1$. Then the upper left block (4.49) admits a CR factorization if and only if the parametrization of the shell mid-surface is isothermal.*

The sufficiency of an isothermal parametrization for the CR factorization was already shown. The conclusion in the opposite direction is quite technical and requires several case distinctions. Therefore, the following lemma is needed.

Lemma 4.9.2. *The assertion that A can be represented in the form:*

$$A = M_4 \begin{pmatrix} \partial_1 & \partial_2 \\ -\partial_2 & \partial_1 \end{pmatrix} M_5 \begin{pmatrix} \partial_1 & -\partial_2 \\ \partial_2 & \partial_1 \end{pmatrix} M_6 + \mathcal{O}\left(\partial^1\right) \tag{4.60}$$

is equivalent to the Definition 4.9.1 in following sense. The representations (4.59) and (4.60) can be transformed into each other if the components of the eigenvectors of M_2 are differentiable functions. This again, is equivalent to the differentability of the eigenvectors of M_5.

4.10 Strain-Free Deformations

The set of all displacement fields $\mathbf{v} = v_j \mathbf{a}^j$ with vanishing strain tensors $\alpha_{(\alpha\beta)}$ and $\omega_{(\alpha\beta)}$ form a linear vector space, the *space of strain-free deformations*. For its characterization, let us consider the time-dependent deformation

$$\mathbf{r}(x) \longmapsto \mathbf{r}(x) + t\mathbf{v}(x) \tag{4.61}$$

of the mid-surface. The derivative with respect to t at the point $t = 0$ is denoted by a dot. The derivative of the first fundamental tensor is

$$\dot{a}_{\alpha\beta} = v_{\alpha|\beta} + v_{\beta|\alpha} - 2b_{\alpha\beta}v_3 = 2\alpha_{(\alpha\beta)}$$

and the derivative of the second fundamental tensor is

$$\dot{b}_{\alpha\beta} = v_{3|\alpha\beta} + v_{\lambda|\beta}b_\alpha^\lambda + v_{\lambda|\alpha}b_\alpha^\lambda + v_\lambda b_{\alpha|\beta}^\lambda - v_3 b_{\alpha\lambda}b_\beta^\lambda. \tag{4.62}$$

The last expression is called the *change of curvature tensor*. It is used in several shell models (e. g. in Koiter's and Naghdi's model) as a strain measure instead of (4.7).

The equations

$$\dot{a}_{\alpha\beta} = \dot{b}_{\alpha\beta} = 0 \tag{4.63}$$

mean that the first and the second fundamental tensor of the deformed mid-surface coincides with the corresponding tensors of the original mid-surface up to terms of order t^2. An exact coincidence implies by virtue of Bonnet's theorem, that both surfaces are identical up to rotations and translations. Therefore, the space of strain-free deformations of Koiter's model, which consists of all solutions of (4.63), is six-dimensional and has a simple structure, described in [10, Lemma 2.5].

The space of strain-free deformations for the DV model has a more complicated structure; and the dimension depends on the geometry of the mid-surface. It consists of all bendings of first order[1] fulfilling the three side conditions $v_{3|\alpha\beta} = 0$. This space can be explicitly described for some special geometrical forms of the mid-surface, e. g., for circular cylinders. In general the problem, under which conditions a bending exists, is rather difficult by itself and requires the distinction of several cases. Such investigations are performed in great detail in [31] and [129].

In the context considered here we are only interested in the question, under which conditions the occurrence of strain-free solutions can be avoided so that the unique solvability of the shell equations can be guaranteed. Since the parametric functions $X_j(x)$ and all terms derived from them are real-analytic, one of the following two cases is valid:

Case 1: The Gaussian curvature $\mathcal{K}$ vanishes identically; or

Case 2: $\mathcal{K}$ vanishes at most on a set of measure zero.

To ensure the non-existence of strain-free solutions, it suffices to require the following six conditions for an arbitrary fixed point $x_0 \in \Omega$:

$$v_j(x_0) = 0, \qquad j = 1, 2, 3, \tag{4.64}$$
$$v_{3,\alpha} = 0, \qquad \alpha = 1, 2, \tag{4.65}$$
$$v_{1,2}(x_0) - v_{2,1}(x_0) = 0. \tag{4.66}$$

This means geometrically, that neither a translation nor a rotation occurs in the vicinity of x_0.

---

[1] In differential geometry, a deformation (4.61) is called a bending of nth order, if, for $t = 0$, all derivatives of $a_{\alpha\beta}$ with respect to t up to the order n vanish [31].

Theorem 4.10.1. *The space of strain-free deformations for the DV model has at most the dimension 6. The system of the six differential equations*

$$\alpha_{(\alpha\beta)} = \omega_{(\alpha\beta)} = 0 \tag{4.67}$$

has only the trivial solution $\mathbf{v} = 0$ *under the additional conditions* (4.64)–(4.66).

Proof. First we remark that the equations (4.64)–(4.67) keep invariant if the parametrization of the mid-surface is changed.

Case 1: We take a parametrization for which all Christoffel symbols vanish. Such a parametrization exists (at least locally) since a surface with a vanishing Gaussian curvature is evolvable (locally) into the plane. Then the relations

$$\omega_{(\alpha\beta)} = -v_{3|\alpha\beta} = -v_{3,\alpha\beta} = 0$$

imply that v_3 is a linear function in both parameters. The additional conditions $v_3(x_0) = v_{3,\alpha}(x_0) = 0$ admit only the solution $v_3 \equiv 0$.

Thus the equations $\alpha_{(\alpha\beta)} = 0$ lead to

$$v_{1,1} = v_{2,2} = v_{1,2} + v_{2,1} = 0,$$

which has only linear solutions

$$v_1 = c_0 x_1 + c_1, \qquad v_2 = -c_0 x_2 + c_2.$$

The constants c_0, c_1, c_2 must vanish due to conditions (4.64) and (4.66).

Case 2: As can be seen from

$$\left(v_{3,\alpha}\mathbf{a}^\alpha\right)_{,\beta} = v_{3|\alpha\beta}\mathbf{a}^\alpha + v_{3,\alpha}b_\beta^\alpha\mathbf{a}^3,$$

the relation $\omega_{(\alpha\beta)} = 0$ implies

$$0 = \varepsilon^{\lambda\beta}\left(v_{3,\alpha}\mathbf{a}^\alpha\right)_{,\beta\lambda} = \varepsilon^{\lambda\beta}\left(v_{3,\alpha}b_\beta^\alpha\mathbf{a}^3\right)_{,\lambda} = \varepsilon^{\lambda\beta}\left[\left(v_{3,\alpha}b_\beta^\alpha\right)_{,\lambda}\mathbf{a}^3 - v_{3,\alpha}b_\beta^\alpha b_\lambda^\gamma\mathbf{a}_\gamma\right].$$

Due to $\varepsilon^{\lambda\beta}b_\beta^\alpha b_\lambda^\gamma = \mathcal{K}\varepsilon^{\gamma\alpha}$ (see (4.17)), we obtain

$$\mathcal{K}v_{3,1} = \mathcal{K}v_{3,2} = 0, \tag{4.68}$$

i. e., v_3 is constant. The assumptions $v_3(x_0) = 0$ lead to $v_3 \equiv 0$ also in this case.

We take an isothermal parametrization. It must be real-analytic as well. Due to (4.4) and (4.56), the strain tensors (4.6) can now be written in the form:

$$\alpha_{(11)} = v_{1|1} = v_{1,1} - \Lambda_{,1}v_1 + \Lambda_{,2}v_2, \tag{4.69}$$

$$\alpha_{(22)} = v_{2|2} = v_{2,2} + \Lambda_{,1}v_1 - \Lambda_{,2}v_2, \tag{4.70}$$

$$\alpha_{(12)} = \frac{1}{2}\left(v_{1|2} + v_{2|1}\right) = \frac{\sqrt{a}}{2}\left[\left(\frac{v_1}{\sqrt{a}}\right)_{,2} + \left(\frac{v_2}{\sqrt{a}}\right)_{,1}\right]. \tag{4.71}$$

From $\alpha_{(12)} = 0$ follows the existence of a function Φ such that $v_1 = \sqrt{a}\,\Phi_{,1}$, $v_2 = -\sqrt{a}\,\Phi_{,2}$. Subtracting (4.70) from (4.69) gives

$$\alpha_{(11)} - \alpha_{(22)} = \sqrt{a}\,\Delta\Phi = 0. \tag{4.72}$$

Φ is a solution of Laplace's equation and hence, it is real-analytic in Ω.

Addition of (4.69) and (4.70) gives

$$\alpha_{(11)} + \alpha_{(22)} = v_{1,1} + v_{2,2} = \sqrt{a}\,(\Phi_{,11} - \Phi_{,22}) + \left(\sqrt{a}\right)_{,1}\Phi_{,1} - \left(\sqrt{a}\right)_{,2}\Phi_{,2} = 0. \tag{4.73}$$

Condition (4.64) implies immediately that the first derivatives of Φ vanish in the point x_0. Further, the equations (4.72), (4.73) and

$$v_{1,2} - v_{2,1} = 2\sqrt{a}\,\Phi_{,12} + \left(\sqrt{a}\right)_{,1}\Phi_{,1} + \left(\sqrt{a}\right)_{,2}\Phi_{,2}$$

admit the conclusion that all second derivatives of Φ vanish as well. By partial differentiation of (4.69) and (4.70) we deduce step by step that all derivatives of Φ vanish at the point x_0. Since Φ is analytic, it must therefore be constant and the displacement field must vanish identically, q. e. d. ∎

4.11 The Energy Bilinear Form

Let $\mathbf{u} = u_j \mathbf{a}^j$, $\mathbf{v} = v_j \mathbf{a}^j$ be two displacement fields and $\alpha_{(\alpha\beta)}[u]$, $\alpha_{(\alpha\beta)}[v]$, $\omega_{(\alpha\beta)}[u]$, $\omega_{(\alpha\beta)}[v]$ the corresponding strain tensors. The integral

$$E(u,v) := \int_{\Omega} H^{\alpha\beta\lambda\mu} \left\{ D\alpha_{(\alpha\beta)}[u]\,\alpha_{(\lambda\mu)}[v] + B\omega_{(\alpha\beta)}[u]\,\omega_{(\lambda\mu)}[v] \right\} \sqrt{a}\,dx_1 dx_2 \tag{4.74}$$

is called the *energy bilinear form*. It makes sense if we assume that both tripels $u := (u_1, u_2, u_3)^\top$ and $v := (v_1, v_2, v_3)^\top$ belong to the Sobolev space

$$H^{(1,2)}(\Omega) := H^1(\Omega) \times H^1(\Omega) \times H^2(\Omega). \tag{4.75}$$

For the following theorem the assumption suffices, that the parametrization (4.2) is 3 times continuously differentiable in $\overline{\Omega}$. Then, the Christoffel symbols and the components of the second fundamental tensor are continuously differentiable functions. This implies the continuity of the bilinear form (4.74) with respect to the norm

$$\|v\|_{(1,2)} := \left(\|v_1\|_1^2 + \|v_2\|_1^2 + \|v_3\|_2^2 \right)^{1/2}. \tag{4.76}$$

Theorem 4.11.1. *Let* $\mathbf{V}$ *be a subspace of* $H^{(1,2)}$ *which has only the intersection zero with the space of strain-free deformations. Then the bilinear form (4.74) is coercive, i. e., there exists a constant* $C > 0$ *with*

$$E(v,v) \geq C \left\| v;\, H^{(1,2)} \right\|^2, \qquad \forall v \in \mathbf{V}. \tag{4.77}$$

A similar result for Koiter's model can be found in [10, Theorem 2.1]. The proof given there can be carried over straightforwardly to the actual case. It becomes even simpler, since $\omega_{(\alpha\beta)}$ enters instead of the change of curvature tensor (4.62) into the energy biliner form. Moreover, the implication $\alpha_{(\alpha\beta)} = \omega_{(\alpha\beta)} = 0 \Rightarrow v = 0$ is already contained in the assumptions here.

Together with Green's formula (see section 6.2) and the theorem of Lax and Milgram, Theorem 4.11.1 gives existence and uniqueness results for solutions of the shell equations. Moreover, convergence results for finite element methods can be derived from (4.77).

5. Levi Functions for the Shell Equations

We prove the ellipticity of the system (4.48) and apply the construction method for Levi functions from chapter 2 to this special system. The space dimension is 2 and thus, integrations in the complex plane can be used for the calculations.

Especially the sections 5.5 and 5.6 are of importance for the algebraic part of the BDIM, i. e., these formulas enter into the computer algebraic routines for symbolic calculation of the Levi functions.

5.1 Ellipticity of the Operator

The shell equations (4.48) are a system of the form (2.1) with the special values n (= space dimension) = 2, m (= number of equations) = 3. The orders of the particular matrix entries fulfill the estimate (2.3) with

$$(s_1, s_2, s_3) = (t_1, t_2, t_3) = (1, 1, 2).$$

Since the operators $A^{\alpha 3}$ and $A^{3\alpha}$ have the order 1, they don't contribute to the principal part (2.4). The principal part $A_0(x, \xi)$ consists of the upper left block

$$A_0^\square(x, \xi) := \frac{D}{\sqrt{a}} \left[\underline{M} \begin{pmatrix} \xi_1 \\ \xi_2 \end{pmatrix} \begin{pmatrix} \xi_1 \\ \xi_2 \end{pmatrix}^\top \underline{M} + \tilde{\nu} \begin{pmatrix} -\xi_2 \\ \xi_1 \end{pmatrix} \begin{pmatrix} -\xi_2 \\ \xi_1 \end{pmatrix}^\top \right]$$

(cmp. (4.49)) and the lower right entry

$$A_0^{33}(x, \xi) = -\frac{B}{\sqrt{a}} \left[\begin{pmatrix} \xi_1 \\ \xi_2 \end{pmatrix}^\top \underline{M} \begin{pmatrix} \xi_1 \\ \xi_2 \end{pmatrix} \right]^2$$

(cmp (4.37) and (4.52)). The scalar a and the matrix $\underline{M}$ depend on x. The constant $\tilde{\nu}$ is defined by (4.53).

To simplify the notation, we set $\xi^\alpha := a^{\alpha\beta} \xi_\beta$. Then

$$\underline{M} \begin{pmatrix} \xi_1 \\ \xi_2 \end{pmatrix} = \sqrt{a} \begin{pmatrix} \xi^1 \\ \xi^2 \end{pmatrix},$$

$$\underline{M}^{-1} \begin{pmatrix} -\xi_2 \\ \xi_1 \end{pmatrix} = \sqrt{a} \begin{pmatrix} a^{22} & -a^{12} \\ -a^{21} & a^{11} \end{pmatrix} \begin{pmatrix} -\xi_2 \\ \xi_1 \end{pmatrix} = \sqrt{a} \begin{pmatrix} -\xi^2 \\ \xi^1 \end{pmatrix},$$

$$A_0^\square(x,\xi) = D\underline{M}\left[\begin{pmatrix}\xi_1\\\xi_2\end{pmatrix}\begin{pmatrix}\xi^1\\\xi^2\end{pmatrix}^{\mathsf{T}} + \tilde{\nu}\begin{pmatrix}-\xi^2\\\xi^1\end{pmatrix}\begin{pmatrix}-\xi_2\\\xi_1\end{pmatrix}^{\mathsf{T}}\right], \quad (5.1)$$

$$A_0^{33}(x,\xi) = -B\sqrt{a}\,(\xi_\alpha\xi^\alpha)^2.$$

Taking into account

$$\det\underline{M} = 1,$$
$$\det A_0^\square(x,\xi) = D^2\left[(\xi_1\xi^1 + \tilde{\nu}\xi_2\xi^2)(\xi_2\xi^2 + \tilde{\nu}\xi_1\xi^1) - \right.$$
$$\left. - (\xi_1\xi^2 - \tilde{\nu}\xi_1\xi^2)(\xi_2\xi^1 - \tilde{\nu}\xi_2\xi^1)\right]$$
$$= D^2\tilde{\nu}\,(\xi_\alpha\xi^\alpha)^2, \quad (5.2)$$

the determinant (2.5) takes the value

$$H(x,\xi) = \det A_0(x,\xi) = -D^2B\tilde{\nu}\sqrt{a(x)}\left[a^{\alpha\beta}(x)\xi_\alpha\xi_\beta\right]^4. \quad (5.3)$$

Note that the first fundamental tensor of a non-degenerated surface is positive definite. The other factors in (5.3) don't vanish as well. Thus, the considered system (4.48) is elliptic in the sense of Douglis-Nirenberg and has the order 8.

Multiplying the matrix (5.1) from the left with $\underline{M}^{-1}$ and calculating the sum of the main diagonal entries, gives the expression

$$D(1+\tilde{\nu})\,(\xi_\alpha\xi^\alpha)$$

which is always positive for $\xi \neq 0$, since $\tilde{\nu} \geq 1/4$. Together with (5.2) we conclude that the matrix $A_0^\square(x,\xi)$ has only positive eigenvalues for $\xi \neq 0$. They can be estimated from below by $DC(\underline{M},\nu)|\xi|^2$, where $C(\underline{M},\nu)$ is a constant, independent of x. It depends only on Poisson's ratio ν and the smallest eigenvalue of the first fundamental tensor in $\overline{\Omega}$. Thus we have

$$-\sum_{j,k=1}^{3} A_0^{jk}(x,i\xi)\eta_j\overline{\eta_k} = -A_0^{\alpha\beta}(x,i\xi)\eta_\alpha\overline{\eta_\beta} - A_0^{33}(x,i\xi)\,|\eta_3|^2$$

$$\geq DC(\underline{M},\nu)|\xi|^2\left[(\eta_1)^2 + (\eta_2)^2\right] + B\sqrt{a}\,(\xi_\alpha\xi^\alpha)^2\,|\eta_3|^2$$

and the condition (2.7) of strong ellipticity is fulfilled with the angle $\phi = \pi$.

5.2 Splitting of the Operator into Principal and Remaining Part

Similar to (2.10), the operator $A = A(x,\partial)$ is split into the difference $A = A_0 - T$. The principal part A_0 has following components:

$$A_0^\square := \begin{pmatrix} A_0^{11} & A_0^{12} \\ A_0^{21} & A_0^{22} \end{pmatrix} \tag{5.4}$$

$$= D\left[\underline{M}(y) \begin{pmatrix} \partial_1 \\ \partial_2 \end{pmatrix} \frac{1}{\sqrt{a(x)}} \begin{pmatrix} \partial_1 \\ \partial_2 \end{pmatrix}^{\mathsf{T}} \underline{M}(y) + \right.$$

$$\left. + \tilde{\nu} \begin{pmatrix} -\partial_2 \\ \partial_1 \end{pmatrix} \frac{1}{\sqrt{a(x)}} \begin{pmatrix} -\partial_2 \\ \partial_1 \end{pmatrix}^{\mathsf{T}} \right],$$

$$A_0^{13} = A_0^{23} = A_0^{31} = A_0^{32} = 0, \tag{5.5}$$

$$A_0^{33} = -B\tilde{\Delta}\frac{1}{\sqrt{a(x)}}\tilde{\Delta}. \tag{5.6}$$

Here we have used the abbreviation

$$\tilde{\Delta} := \begin{pmatrix} \partial_1 \\ \partial_2 \end{pmatrix}^{\mathsf{T}} \underline{M}(y) \begin{pmatrix} \partial_1 \\ \partial_2 \end{pmatrix}.$$

Note that the construction of A_0 according to (5.4)–(5.6) does not exactly coincide with the requirement $A_0 = A_0(y, \partial)$ since the function $1/\sqrt{a(x)}$ also should have then been froozen at the point y. The construction of a right inverse to A_0 becomes simpler however, if this function remains unchanged. For the components of T we obtain:

$$T^\square := \begin{pmatrix} T^{11} & T^{12} \\ T^{21} & T^{22} \end{pmatrix} \tag{5.7}$$

$$= D\left\{ \underline{M}(x) \begin{pmatrix} \partial_1 \\ \partial_2 \end{pmatrix} \frac{1}{\sqrt{a(x)}} \begin{pmatrix} \partial_1 \\ \partial_2 \end{pmatrix}^{\mathsf{T}} [\underline{M}(y) - \underline{M}(x)] + \right.$$

$$\left. + [\underline{M}(y) - \underline{M}(x)] \begin{pmatrix} \partial_1 \\ \partial_2 \end{pmatrix} \frac{1}{\sqrt{a(x)}} \begin{pmatrix} \partial_1 \\ \partial_2 \end{pmatrix}^{\mathsf{T}} \underline{M}(y) - 2\tilde{\nu}\mathcal{K}\underline{M}(x) \right\}$$

$$T_K := \begin{pmatrix} T^{13} \\ T^{23} \end{pmatrix} = 2D\,\underline{M}(x)\left[\begin{pmatrix} \partial_1 \\ \partial_2 \end{pmatrix} \mathcal{H}(x) - \tilde{\nu}\,\underline{N}(x) \begin{pmatrix} \partial_1 \\ \partial_2 \end{pmatrix} \right]$$

$$T_K^{\mathsf{T}} := \begin{pmatrix} T^{31} \\ T^{32} \end{pmatrix}^{\mathsf{T}} = 2D\left[-\mathcal{H}(x) \begin{pmatrix} \partial_1 \\ \partial_2 \end{pmatrix}^{\mathsf{T}} + \tilde{\nu} \begin{pmatrix} \partial_1 \\ \partial_2 \end{pmatrix}^{\mathsf{T}} \underline{N}^{\mathsf{T}}(x) \right] \underline{M}(x)$$

$$T^{33} = B\left[\Delta_M \frac{1}{\sqrt{a(x)}} \left(\Delta_M - \tilde{\Delta} \right) + \left(\Delta_M - \tilde{\Delta} \right) \frac{1}{\sqrt{a(x)}} \tilde{\Delta} + \right. \tag{5.8}$$

$$\left. + 2\tilde{\nu} \begin{pmatrix} \partial_1 \\ \partial_2 \end{pmatrix}^{\mathsf{T}} \mathcal{K}(x)\underline{M}(x) \begin{pmatrix} \partial_1 \\ \partial_2 \end{pmatrix} \right] + 4D\sqrt{a(x)}\left[\mathcal{H}^2(x) - \tilde{\nu}\mathcal{K}(x) \right]$$

The operator $A_0^{(-1)}T$ used in the iteration step (2.15) has the block-matrix representation

$$A_0^{(-1)} T = \begin{pmatrix} \left(A_0^{\square}\right)^{(-1)} T^{\square} & \left(A_0^{\square}\right)^{(-1)} T_K \\ \left(A_0^{33}\right)^{(-1)} T_K^{\top} & \left(A_0^{33}\right)^{(-1)} T^{33} \end{pmatrix}, \tag{5.9}$$

where $\left(A_0^{\square}\right)^{(-1)}$ and $\left(A_0^{33}\right)^{(-1)}$ are right inverses to the operators (5.4) and (5.6), respectively.

5.3 Local Isothermal Parametrization

The components of the distance vector $\vartheta = x - y$ are equipped with upper indices:

$$\vartheta^1 := x_1 - y_1, \qquad \vartheta^2 := x_2 - y_2. \tag{5.10}$$

Then, the summation convention can be applied to form the expression

$$\tilde{\rho} := \sqrt{a_{\alpha\beta}(y)\vartheta^\alpha\vartheta^\beta}. \tag{5.11}$$

As can be seen from the Taylor expansion

$$\mathbf{r}(x) = \mathbf{r}(y) + \vartheta^\alpha \mathbf{a}_\alpha + \mathcal{O}\left(|\vartheta|^2\right),$$

$\tilde{\rho}$ coincides (in first approximation) with the distance of the points $\mathbf{r}(x)$ and $\mathbf{r}(y)$ on the mid-surface if x and y are close to each other.

Let $J = J(y)$ be a 2×2 matrix which fulfills the equation

$$J^\top J = \underline{M}(y). \tag{5.12}$$

For instance, one can choose

$$J = \begin{pmatrix} J_{11}(y) & J_{12}(y) \\ 0 & J_{22}(y) \end{pmatrix} \tag{5.13}$$

with

$$J_{11}(y) := \sqrt{a^{11}(y)}\sqrt{a(y)}, \quad J_{12}(y) := a^{12}(y)\sqrt{\frac{\sqrt{a(y)}}{a^{11}(y)}}, \quad J_{22}(y) := \frac{1}{J_{11}(y)}.$$

Each other matrix obtained by multiplication of (5.13) by an arbitrary orthogonal matrix from the left, fulfills (5.12) and can be used as well for the following substitution of variables:

$$\begin{pmatrix} \theta_1 \\ \theta_2 \end{pmatrix} := J^{-\top} \begin{pmatrix} \vartheta^1 \\ \vartheta^2 \end{pmatrix}. \tag{5.14}$$

The new variables $\theta = (\theta_1, \theta_2)$ can be interpreted as *local isothermal parameters* of the mid-surface. This means that the parametrization with respect to θ is isothermal in the neighbourhood of the point y in first approximation. Thereby holds

$$\sqrt{\theta_1^2 + \theta_2^2} = \frac{\tilde{\rho}}{\sqrt[4]{a(y)}} =: \rho. \tag{5.15}$$

Further, from (5.14) and $\det J = 1$ there follows

$$\begin{pmatrix} -\theta_2 \\ \theta_1 \end{pmatrix} := J \begin{pmatrix} -\vartheta^2 \\ \vartheta^1 \end{pmatrix}. \tag{5.16}$$

For the partial derivatives with respect to θ_α, denoted by $\tilde{\partial}_\alpha$, the following relations are valid:

$$\begin{pmatrix} \tilde{\partial}_1 \\ \tilde{\partial}_2 \end{pmatrix} = J \begin{pmatrix} \partial_1 \\ \partial_2 \end{pmatrix}, \tag{5.17}$$

$$\begin{pmatrix} -\partial_2 \\ \partial_1 \end{pmatrix} = J^\mathsf{T} \begin{pmatrix} -\tilde{\partial}_2 \\ \tilde{\partial}_1 \end{pmatrix}. \tag{5.18}$$

The upper left block of A_0 admits the factorization

$$\begin{aligned}
A_0^\square &= D \left[J^\mathsf{T} \begin{pmatrix} \tilde{\partial}_1 \\ \tilde{\partial}_2 \end{pmatrix} \frac{1}{\sqrt{a(x)}} \begin{pmatrix} \tilde{\partial}_1 \\ \tilde{\partial}_2 \end{pmatrix}^\mathsf{T} J + \right. \\
&\quad \left. + \tilde{\nu} J^\mathsf{T} \begin{pmatrix} -\tilde{\partial}_2 \\ \tilde{\partial}_1 \end{pmatrix} \frac{1}{\sqrt{a(x)}} \begin{pmatrix} -\tilde{\partial}_2 \\ \tilde{\partial}_1 \end{pmatrix}^\mathsf{T} J \right] \\
&= D J^\mathsf{T} \begin{pmatrix} \tilde{\partial}_1 & -\tilde{\partial}_2 \\ \tilde{\partial}_2 & \tilde{\partial}_1 \end{pmatrix} \frac{1}{\sqrt{a(x)}} \begin{pmatrix} 1 & 0 \\ 0 & \tilde{\nu} \end{pmatrix} \begin{pmatrix} \tilde{\partial}_1 & \tilde{\partial}_2 \\ -\tilde{\partial}_2 & \tilde{\partial}_1 \end{pmatrix} J. \tag{5.19}
\end{aligned}$$

Of course, it is possible as well to use the "correct" isothermal parameters instead of the local isothermal ones. Due to $\underline{M}(x) - \underline{M}(y) = 0$ we would have essential simplifications then for the operators (5.7) and (5.8), as already mentioned in section 4.8. This seems to be recommendable, however, only if the isothermal parametrization is explicitly known. Otherwise it is only possible to construct a series expansion for the isothermal parameters in dependence on the original ones. This must be done for each point y separately. The complete procedure becomes more complicated and the expense to calculate Levi functions of higher order increases.

5.4 Pseudohomogeneous Functions of Special Structure

In section 1.8 we have introduced the classes $\Psi(\mu, y)$ and $\Psi^\bullet(\mu, y)$, which the particular pseudohomogeneous terms of the Levi functions belong to. Since we restrict now to the special case of the shell equations (4.48), it is possible to describe the structure of the occuring pseudohomogeneous functions more precisely.

For $n = 2$, the set of all homogeneous polynomials of degree μ (introduced in section 1.1), has the form

$$\text{Hom}^{(\mu)} := \left\{ \sum_{j=0}^{\mu} h_j\, \theta_1^{\mu-j} \theta_2^j \;\Big|\; h_0, \dots, h_\mu \in \mathbb{R} \right\}. \tag{5.20}$$

The coefficients $h_0, \dots, h_\mu$ may depend on y but not on x. For $\mu < 0$ we set $\text{Hom}^{(\mu)} = \{0\}$.

Let λ, μ be two integers with $\lambda \geq 0$, $\mu \geq -2$. We introduce the spaces

$$\mathcal{G}_{\mu,\lambda} := \left\{ p(\theta_1,\theta_2)\ln(\rho) + \frac{q(\theta_1,\theta_2)}{\rho^{2\lambda}} \;\Big|\; p \in \text{Hom}^{(\mu)},\; q \in \text{Hom}^{(\mu+2\lambda)} \right\} \tag{5.21}$$

of pseudohomogeneous functions of degree μ. They are finite-dimensional and have the dimension

$$\dim(\mathcal{G}_{\mu,\lambda}) = 2(\mu + \lambda + 1) \tag{5.22}$$

for $\mu \geq 0$. Obviously, the functions from $\mathcal{G}_{\mu,\lambda}$ fulfill the parity condition from Definition 1.13.1. Thus we have $\mathcal{G}_{\mu,\lambda} \subset \Psi(\mu, y)$ for each λ.

We identify each rational function of the form $q/\rho^{2\lambda}$ occuring in (5.21) with all functions derived from it by multiplying nominator and denominator with the same power of ρ^2. Then the inclusions

$$\mathcal{G}_{\mu,0} \subset \mathcal{G}_{\mu,1} \subset \mathcal{G}_{\mu,2} \subset \dots \mathcal{G}_{\mu,\infty}$$

are valid, with

$$\mathcal{G}_{\mu,\infty} := \bigcup_{\lambda=0}^{\infty} \mathcal{G}_{\mu,\lambda}.$$

The series $\Psi(\geq \mu, y)$, introduced in Definition 1.13.1, are now specified as follows:

Definition 5.4.1. *Let $\mu \geq -2$ be an integer and let $\chi = \{\chi_0, \chi_1, \chi_2, \dots\}$ be a sequence of natural numbers. If a constant $\varepsilon > 0$ existst such that the series*

$$f(x) = \sum_{j=\mu}^{\infty} f_j(x), \qquad f_j \in \mathcal{G}_{j,\chi_{j-\mu}} \tag{5.23}$$

converges uniformly for all $\varepsilon' < \varepsilon$ in the circular ring $\mathcal{K}_\varepsilon \setminus \mathcal{K}_{\varepsilon'}$, then we write $f \in \mathcal{U}[\mu, \chi]$.

The radius of convergence is defined as in Definition 1.11.1.

Clearly, the inclusion

$$\mathcal{U}[\mu, \chi] \subset \Psi(\geq \mu, y)$$

holds, independent of the sequence χ.

Let $\mu, k \geq -2$ be two integers and χ, σ be two sequences of natural numbers. For each pair $f \in \mathcal{U}[\mu, \chi]$, $g \in \mathcal{U}[k, \sigma]$, the sum $f + g$ is a locally convergent series, where the radius of convergence coincides with the minimal one of both summands. For the sum of the corresponding vector spaces holds the relation:

$$\mathcal{U}[\mu, \chi] \oplus \mathcal{U}[k, \sigma] = \begin{cases} \mathcal{U}\left[k, \max\left(V^{\mu-k}\chi, \sigma\right)\right], & \text{for } \mu \geq k, \\ \mathcal{U}\left[\mu, \max\left(\chi, V^{k-\mu}\sigma\right)\right], & \text{for } \mu \leq k, \end{cases} \qquad (5.24)$$

where V denotes the shift operator

$$V\chi := \{0, \chi_0, \chi_1, \chi_2, \ldots\}.$$

The maximum and the (later used) sum of two sequences is to be understood elementwise.

5.5 Inversion of the Cauchy-Riemann Operators

The block (5.19) is a product of multiplication operators and Cauchy-Riemann operators. In the iteration scheme (2.14), (2.15), the right inverse of (5.19) is applied to the first two components of the corresponding column vectors. These components contain the δ-distribution or pseudohomogeneous functions of degree ≥ -1. For an efficient calculation of the right inverse, we apply complex methods and introduce the complex variables

$$z := \theta_1 + i\theta_2, \qquad \zeta := \theta_1 - i\theta_2. \qquad (5.25)$$

Each column vector $[f_1(\theta_1, \theta_2), f_2(\theta_1, \theta_2)]^\top$ of real-valued functions is identified with the complex-valued function

$$f(z, \zeta) = f_1\left(\frac{z+\zeta}{2}, \frac{i(\zeta-z)}{2}\right) + if_2\left(\frac{z+\zeta}{2}, \frac{i(\zeta-z)}{2}\right). \qquad (5.26)$$

Then the Cauchy-Riemann operators can be written as partial derivatives with respect to z and ζ:

$$\frac{1}{2}\begin{pmatrix} \tilde{\partial}_1 & \tilde{\partial}_2 \\ -\tilde{\partial}_2 & \tilde{\partial}_1 \end{pmatrix}\begin{pmatrix} f_1(\theta_1, \theta_2) \\ f_2(\theta_1, \theta_2) \end{pmatrix} = \begin{pmatrix} \mathfrak{Re} \\ \mathfrak{Im} \end{pmatrix}\frac{\partial}{\partial z} f(z, \zeta), \qquad (5.27)$$

$$\frac{1}{2}\begin{pmatrix} \tilde{\partial}_1 & -\tilde{\partial}_2 \\ \tilde{\partial}_2 & \tilde{\partial}_1 \end{pmatrix}\begin{pmatrix} f_1(\theta_1, \theta_2) \\ f_2(\theta_1, \theta_2) \end{pmatrix} = \begin{pmatrix} \mathfrak{Re} \\ \mathfrak{Im} \end{pmatrix}\frac{\partial}{\partial \zeta} f(z, \zeta). \qquad (5.28)$$

Right inverses to (5.27) and (5.28) are obtained by integrations in the complex plane:

$$\partial_z^{(-1)} f(z, \zeta) := \int_{z_0}^{z} f(\tau, \zeta)d\tau, \qquad (5.29)$$

$$\partial_\zeta^{(-1)} f(z, \zeta) := \int_{\zeta_0}^{\zeta} f(z, \tau)d\tau. \qquad (5.30)$$

These are the well-known Pompeiu's formulas.
    Suppose

$$f_1, f_2 \in \mathcal{U}[-1, \infty] \subset \Psi(\geq -1, y)$$

and let ε be the minimum of both radii of convergence. Then the function (5.26), after subtraction of all terms of degrees -1 and 0, admits an absolutely convergent series expansion in the zero-neighbourhood

$$\left\{(z,\zeta) \in \mathbb{C}^2 \,\big|: |z| < \varepsilon \text{ and } |\zeta| < \varepsilon\right\}.$$

If the paths of integration in (5.29) and (5.30) don't contain the origin, then the integrals exist in the usual sense. To achieve the desired mapping properties, it is necessary however, to put $z_0 = \zeta_0 = 0$. The divergent parts occuring then, must be interpreted as partie-finie limits for $z_0 \to 0$ or $\zeta_0 \to 0$. Doing so, the right inverses $\partial_z^{(-1)}$ and $\partial_\zeta^{(-1)}$ are well-defined for all pairs of functions $f_1, f_2 \in \mathcal{U}[-1, \infty]$. Moreover, it is possible to evaluate them explicitly. All of the particular pseudohomogeneous parts of the function $f_1 + if_2$ are transformed by the substitutions (5.14) and (5.25) into a linear combination of expressions of the form

$$z^k \zeta^l \ln \rho \qquad \text{and} \qquad \frac{z^k \zeta^l}{\rho^{2j}} \tag{5.31}$$

with $k, l, j \in \mathbb{N}_0$ and $k + l - 2j \geq -1$. Due to (5.15), one has $\rho = |z| = |\zeta|$. Elementary integration formulas can be used to calculate:

$$\partial_z^{(-1)}\left(z^k \zeta^l \ln \rho\right) = \frac{z^{k+1}\zeta^l}{k+1}\left[\ln \rho - \frac{1}{2(k+1)}\right], \tag{5.32}$$

$$\partial_z^{(-1)}\left(\frac{z^k \zeta^l}{\rho^{2j}}\right) = \begin{cases} \frac{z^{k+1}\zeta^l}{(k+1-j)\rho^{2j}}, & \text{if } k+1-j \neq 0, \\ 2\zeta^{l-j} \ln \rho, & \text{if } k+1-j = 0. \end{cases} \tag{5.33}$$

It can be seen that $\partial_z^{(-1)}$ raises the degree of pseudohomogeneity by one step, but the power of ρ in the denominator stays constant. The same is true for $\partial_\zeta^{(-1)}$. Thus we obtain the mapping properties:

$$\left.\begin{array}{ll} \partial_z^{(-1)}, \partial_\zeta^{(-1)}: & (\mathcal{G}_{\mu,\lambda})^2 \longmapsto (\mathcal{G}_{\mu+1,\lambda})^2, \quad \forall \mu \geq -1, \lambda \geq 0, \\ \partial_z^{(-1)}, \partial_\zeta^{(-1)}: & \left(\mathcal{U}[\mu, \chi]\right)^2 \longmapsto \left(\mathcal{U}[\mu+1, \chi]\right)^2, \quad \forall \mu \geq -1, \forall \chi. \end{array}\right\} \tag{5.34}$$

A right inverse to the Laplacian

$$\tilde{\Delta} = \tilde{\partial}_1^2 + \tilde{\partial}_2^2 = 4\partial_z \partial_\zeta$$

can be obtained by superposition of $\partial_z^{(-1)}$ and $\partial_\zeta^{(-1)}$. This gives the following relations for $k, l, j \in \mathbb{N}_0$ and $k + l - 2j > -1$:

$$\tilde{\Delta}^{(-1)}\left(z^k \zeta^l \ln \rho\right) = \frac{z^{k+1}\zeta^{l+1}}{4(k+1)(l+1)}\left(\ln \rho - \frac{1}{2(k+1)} - \frac{1}{2(l+1)}\right), \tag{5.35}$$

$$\tilde{\Delta}^{(-1)}\left(\frac{z^k \zeta^l}{\rho^{2j}}\right) = \begin{cases} \frac{\rho^2 z^k \zeta^l}{4(k+1-j)(l+1-j)\rho^{2j}}, & \text{if} \quad (k+1-j)(l+1-j) \neq 0, \\ \frac{z^{k+1-j}}{2(k+1-j)} \ln \rho, & \text{if} \quad j = l+1 \quad (\Rightarrow k \geq j), \\ \frac{\zeta^{l+1-j}}{2(l+1-j)} \ln \rho, & \text{if} \quad j = k+1 \quad (\Rightarrow l \geq j). \end{cases}$$

As can be seen, $\tilde{\Delta}^{(-1)}$ reduces the power of ρ^{-2} by one step:

$$\left.\begin{array}{ll} \tilde{\Delta}^{(-1)}: & \mathcal{G}_{\mu,\lambda} \longmapsto \mathcal{G}_{\mu+2,\max(\lambda-1,0)}, \qquad \forall \mu \geq -1, \ \lambda \geq 0, \\ \tilde{\Delta}^{(-1)}: & \mathcal{U}[\mu,\chi] \longmapsto \mathcal{U}[\mu+2,\max(\chi-1,0)], \quad \forall \mu \geq -1, \ \forall \chi. \end{array}\right\} \quad (5.36)$$

5.6 The Initial Terms of the Levi Function

Right inverses to the operators (5.19) and (5.6) can be written in the form

$$\left(A_0^\square\right)^{(-1)} = \frac{1}{4D} J^{-1} \partial_z^{(-1)} \begin{pmatrix} 1 & 0 \\ 0 & \tilde{\nu}^{-1} \end{pmatrix} \sqrt{a(x)} \, \partial_\zeta^{(-1)} J^{-\top}, \quad (5.37)$$

$$\left(A_0^{33}\right)^{(-1)} = -\frac{1}{B} \tilde{\Delta}^{(-1)} \sqrt{a(x)} \, \tilde{\Delta}^{(-1)}. \quad (5.38)$$

The complete right inverse $A_0^{(-1)}$ built from (5.37) and (5.38), can be applied now to tripels of functions from the classes (5.21), (5.23) whereby the result can be obtained in closed form. Note that the degree μ, the exponent λ or the sequence χ are allowed to be different for the three functions.

Let us calculate the initial term $N^{(0)} = A_0^{(-1)} \delta(\vartheta) I_{3\times 3}$ of the iteration step (2.14). We make use of the well-known fundamental solution for the Cauchy-Riemann equation

$$\partial_\zeta^{(-1)} \, \delta(\theta) I_{2\times 2} = \frac{1}{\pi \rho^2} \begin{pmatrix} \theta_1 & \theta_2 \\ -\theta_2 & \theta_1 \end{pmatrix} \quad (5.39)$$

and the fundamental solution for the Laplacian

$$\tilde{\Delta}^{(-1)} \delta(\theta) := \frac{1}{2\pi} \ln \rho. \quad (5.40)$$

From $\det J \equiv 1$ follows $\delta(\vartheta) = \delta(\theta)$. Further we use that $\delta \cdot I_{2\times 2}$ commutes with the multiplication operator $J^{-\top}$. This gives

$$N_\square^{(0)} := \begin{pmatrix} N_{11}^{(0)} & N_{12}^{(0)} \\ N_{21}^{(0)} & N_{22}^{(0)} \end{pmatrix} = \left(A_0^\square\right)^{(-1)} \delta(\vartheta) I_{2\times 2}$$

$$= \frac{1}{4\pi D} J^{-1} \partial_z^{(-1)} \begin{pmatrix} 1 & 0 \\ 0 & \tilde{\nu}^{-1} \end{pmatrix} \frac{\sqrt{a(x)}}{\rho^2} \begin{pmatrix} \theta_1 & \theta_2 \\ -\theta_2 & \theta_1 \end{pmatrix} J^{-\top},$$

$$N_{33}^{(0)} = -\frac{1}{B} \tilde{\Delta}^{(-1)} \sqrt{a(x)} \tilde{\Delta}^{(-1)} \delta(\vartheta) =$$

$$= -\frac{1}{2\pi B} \tilde{\Delta}^{(-1)} \sqrt{a(x)} \ln \rho, \quad (5.41)$$

$$N_{\alpha 3}^{(0)} = N_{3\alpha}^{(0)} = 0. \quad (5.42)$$

The upper left block starts with terms of degree 0. They can be written in the form

$$P_0 N_\square^{(0)} = \frac{\sqrt{a(y)}}{8\pi D} J^{-1} \left[(1 + \tilde{\nu}^{-1}) Q_1 + (1 - \tilde{\nu}^{-1}) Q_2 \right] J^{-\top}, \qquad (5.43)$$

with

$$Q_1 := \partial_z^{(-1)} \left[\frac{1}{\rho^2} \begin{pmatrix} \theta_1 & \theta_2 \\ -\theta_2 & \theta_1 \end{pmatrix} \right] = \begin{pmatrix} \Re & -\Im \\ \Im & \Re \end{pmatrix} 2 \ln \rho =$$

$$= 2 \ln \rho \begin{pmatrix} 1 & 0 \\ 0 & 1 \end{pmatrix},$$

$$Q_2 := \partial_z^{(-1)} \left[\frac{1}{\rho^2} \begin{pmatrix} \theta_1 & \theta_2 \\ \theta_2 & -\theta_1 \end{pmatrix} \right] = \begin{pmatrix} \Re & \Im \\ \Im & -\Re \end{pmatrix} \frac{z^2}{\rho^2} =$$

$$= \frac{2}{\rho^2} \begin{pmatrix} \theta_1 \\ \theta_2 \end{pmatrix} \begin{pmatrix} \theta_1 \\ \theta_2 \end{pmatrix}^\top - \begin{pmatrix} 1 & 0 \\ 0 & 1 \end{pmatrix}.$$

Here we have used formula (5.33) for the special cases

$$\partial_z^{(-1)} \frac{\zeta}{\rho^2} = 2 \ln \rho, \qquad \partial_z^{(-1)} \frac{z}{\rho^2} = \frac{z^2}{\rho^2}.$$

Furthermore, relations (5.12) and (5.14) give:

$$J^{-1} Q_1 J^{-\top} = 2 \ln \rho \, \underline{M}^{-1}(y),$$

$$J^{-1} Q_2 J^{-\top} = \underline{M}^{-1}(y) \frac{2}{\rho^2} \begin{pmatrix} \vartheta_1 \\ \vartheta_2 \end{pmatrix} \begin{pmatrix} \vartheta_1 \\ \vartheta_2 \end{pmatrix}^\top \underline{M}^{-1}(y) - \underline{M}^{-1}(y).$$

Now we put everything together. With (5.15) and

$$1 - \tilde{\nu}^{-1} = -\frac{1+\nu}{1-\nu}, \qquad 1 + \tilde{\nu}^{-1} = \frac{3-\nu}{1-\nu},$$

we obtain from (5.43):

$$P_0 N_\square^{(0)} = \frac{\sqrt{a(y)}}{4\pi D(1-\nu)} \left\{ \left[(3-\nu) \ln \rho + \frac{1+\nu}{2} \right] \underline{M}^{-1}(y) - \right.$$

$$\left. - \frac{1+\nu}{\rho^2} \underline{M}^{-1}(y) \begin{pmatrix} (\vartheta^1)^2 & \vartheta^1 \vartheta^2 \\ \vartheta^1 \vartheta^2 & (\vartheta^2)^2 \end{pmatrix} \underline{M}^{-1}(y) \right\}. \qquad (5.44)$$

The constant term $(1+\nu)/2$ in the square bracket can be dropped, since it doesn't contribute to the singular behaviour. For vanishing Gaussian curvature and orthogonal parametrization, (5.44) appears to be the well-known Kelvin's fundamental solution for the two-dimensional equations of elasticity. Recall that

$$\underline{M}^{-1}(y) = \frac{1}{\sqrt{a}} \begin{pmatrix} a_{11} & a_{12} \\ a_{21} & a_{22} \end{pmatrix}$$

is the first fundamental tensor, normalized to the determinant 1.

Finally, let us calculate the first two terms of $N_{33}^{(0)}$. Formula (5.35) is used for the following cases:

$$\tilde{\Delta}^{(-1)} \ln \rho = \frac{\rho^2}{4}(\ln \rho - 1),$$

$$\tilde{\Delta}^{(-1)} \begin{pmatrix} z \\ \zeta \end{pmatrix} \ln \rho = \frac{\rho^2}{8}\left(\ln \rho - \frac{3}{4}\right) \begin{pmatrix} z \\ \zeta \end{pmatrix}.$$

From

$$\begin{pmatrix} \vartheta^1 \\ \vartheta^2 \end{pmatrix} = J^\top \begin{pmatrix} \theta_1 \\ \theta_2 \end{pmatrix} = \frac{1}{2} J^\top \begin{pmatrix} 1 & 1 \\ -i & i \end{pmatrix} \begin{pmatrix} z \\ \zeta \end{pmatrix}$$

we conclude

$$\tilde{\Delta}^{(-1)} \begin{pmatrix} \vartheta^1 \\ \vartheta^2 \end{pmatrix} \ln \rho = \frac{\rho^2}{8}\left(\ln \rho - \frac{3}{4}\right) \begin{pmatrix} \vartheta^1 \\ \vartheta^2 \end{pmatrix}.$$

Inserting this and the Taylor expansion

$$\sqrt{a(x)} = \sqrt{a(y)} \left[1 + \vartheta^\alpha \Gamma_{\alpha\beta}^\beta(y)\right] + \mathcal{O}\left(|\vartheta|^2\right)$$

(cmp. (4.5)) into (5.41), we obtain

$$(P_2 + P_3) N_{33}^{(0)} = -\frac{\sqrt{a(y)}}{8\pi B} \left[\rho^2 (\ln \rho - 1) + \frac{1}{2}\vartheta^\alpha \Gamma_{\alpha\beta}^\beta(y) \rho^2 \left(\ln \rho - \frac{3}{4}\right)\right].$$

$$(5.45)$$

The calculation of higher degree terms of the Levi function is a very tedious work, if done by hand. Symbolic manipulation with Computer algebra offers itself here as a useful tool.

We have used symbolic manipulation to calculate the term $P_3 N_{33}^{(1)}$. After adding it to (5.45), we received for the lower right position of the Levi function the expression

$$(P_2 + P_3) G_{33} = -\frac{\ln \rho}{8\pi B} \left[\sqrt{a(y)} \rho^2 + \Gamma_{\alpha\beta}^\lambda(y) a_{\lambda\gamma}(y)\vartheta^\alpha \vartheta^\beta \vartheta^\gamma\right] \qquad (5.46)$$

$$+ \text{ polynomial terms.}$$

We want to mention an observation from the practical use of Computer algebra. The length of the formulas for the consecutive terms of the Levi function increases exponentially. A serious problem arises from the necessary simplifications of the obtained expressions. Without simplifications, the calculation of $N_{\alpha\beta}^{(3)}$ already exceeded all the resources on our workstation.

Fortunately, a Levi function build from (5.44) and (5.46) is completely sufficient already for numerical purposes (see chapter 7). Though adding the next terms would lead to smoother integral kernels on some positions, the time to calculate the corresponding integral kernels (not to mention the discretization and implementation) increases significantly.

From the viewpoint of numerical expense there exists an optimal cut-off order for the Levi function which is, roughly estimated, quite small. Taking further into account the code developing time, Levi functions of first or second degree seems to be the best choice.

That is why we will not calculate the higher order terms but only specify the structure of these terms. Especially we will determine, which power of ρ^{-2} occurs at each position.

5.7 Mapping Properties of the Individual Operators

As can be seen from the relation

$$\tilde{\partial}_\alpha \left(p\ln(\rho) + \frac{q}{\rho^{2\lambda}} \right) = \left(\tilde{\partial}_\alpha p \right) \ln(\rho) + \frac{\theta_\alpha p}{\rho^2} + \frac{\tilde{\partial}_\alpha q}{\rho^{2\lambda}} - \frac{2\lambda\theta_\alpha q}{\rho^{2\lambda+2}}, \tag{5.47}$$

the partial derivatives with respect to θ_α have the property:

$$\tilde{\partial}_\alpha : \mathcal{G}_{\mu,\lambda} \longmapsto \mathcal{G}_{\mu-1,\lambda+1}. \tag{5.48}$$

For the second derivatives therefore there holds:

$$\tilde{\partial}_\alpha^2 : \mathcal{G}_{\mu,\lambda} \longmapsto \mathcal{G}_{\mu-2,\lambda+2}.$$

The Laplacian, however, raises the power of ρ^{-2} only be one step:

$$\begin{pmatrix} \partial_1 \\ \partial_2 \end{pmatrix}^\top \underline{M}(x) \begin{pmatrix} \partial_1 \\ \partial_2 \end{pmatrix} = \tilde{\Delta} : \mathcal{G}_{\mu,\lambda} \longmapsto \mathcal{G}_{\mu-2,\lambda+1}. \tag{5.49}$$

This can be seen from

$$\tilde{\Delta} \left(p\ln(\rho) + \frac{q}{\rho^{2\lambda}} \right) = \left(\tilde{\Delta}p \right) \ln(\rho) + 2\frac{\theta_1(\tilde{\partial}_1 p) + \theta_2(\tilde{\partial}_2 p)}{\rho^2} + \frac{\tilde{\Delta}q}{\rho^{2\lambda}} +$$
$$+ 4\lambda^2 \frac{q}{\rho^{2\lambda+2}} - 4\lambda\frac{\theta_1(\tilde{\partial}_1 q) + \theta_2(\tilde{\partial}_2 q)}{\rho^{2\lambda+2}}. \tag{5.50}$$

To simplify the notation in what follows, a constant sequence of numbers $\{k, k, k, \ldots\}$ is denoted by the number k itself.

Let $g(x)$ be a C^∞ function and let its Taylor expansion about the point $x = y$ have a positive radius of convergence. The operator of multiplication with g maps as follows:

$$g : \mathcal{G}_{\mu,\lambda} \longmapsto \mathcal{U}[\mu, \lambda]. \tag{5.51}$$

If the Taylor series starts with linear terms, which it valid, e. g., for the components of $\underline{M}(x) - \underline{M}(y)$, then the degree of homogeneity is raised by one step.

The mapping properties (5.48)–(5.51) can be carried over to series from the classes $\mathcal{U}[\mu, \chi]$:

$$\tilde{\partial}_\alpha : \mathcal{U}[\mu, \chi] \longmapsto \mathcal{U}[\mu - 1, \chi + 1], \qquad \tilde{\Delta}^{(-1)} : \mathcal{U}[\mu, \chi] \longmapsto \mathcal{U}[\mu - 2, \chi + 1],$$

and so on. After the properties of the particular operators are clarified, the behaviour of the components of the operator T, given by (5.7)–(5.8), can be investigated. For convenience we still introduce the following notation: $(\mathcal{M})^{\beta \times \alpha}$ denotes the set of all $\beta \times \alpha$ matries, the entries of which belong to $\mathcal{M}$. The indices α, β take the values 1 and 2.

As can be seen from (5.48)–(5.51), for all $\mu \geq 0$ and all sequences χ holds:

$$T^\square \quad : \quad \left(\mathcal{U}[\mu, \chi] \right)^{2 \times \alpha} \longmapsto \left(\mathcal{U}[\mu - 1, \chi + 2] \right)^{2 \times \alpha}, \tag{5.52}$$

$$T_K \quad : \quad \left(\mathcal{U}[\mu, \chi] \right)^{1 \times \alpha} \longmapsto \left(\mathcal{U}[\mu - 1, \chi + 1] \right)^{2 \times \alpha}, \tag{5.53}$$

$$T_K^\top \quad : \quad \left(\mathcal{U}[\mu, \chi] \right)^{2 \times \alpha} \longmapsto \left(\mathcal{U}[\mu - 1, \chi + 1] \right)^{1 \times \alpha}. \tag{5.54}$$

The relations are slightly more complicated for the operator T^{33}. From

$$\Delta_M - \tilde{\Delta} = \begin{pmatrix} \partial_1 \\ \partial_2 \end{pmatrix}^\top [\underline{M}(x) - \underline{M}(y)] \begin{pmatrix} \partial_1 \\ \partial_2 \end{pmatrix} \quad : \quad \mathcal{G}_{\mu, \lambda} \longmapsto \mathcal{U}[\mu - 1, \lambda + 2]$$

and (5.24), (5.49) we obtain:

$$\Delta_M \quad : \quad \mathcal{G}_{\mu, \lambda} \longmapsto \mathcal{U}[\mu - 2, \lambda + 1] \oplus \mathcal{U}[\mu - 1, \lambda + 2] = \mathcal{U}[\mu - 2, \lambda + 1 + V(1)].$$

The obtained sequence has the form

$$\lambda + 1 + V(1) = \max [\lambda + 1, V(\lambda + 2)] = \lambda + \{1, 2, 2, 2, \ldots\}.$$

Evaluating all particular summands on the right-hand side of (5.8) shows:

$$T^{33} \quad : \quad \mathcal{G}_{\mu, \lambda} \longmapsto \mathcal{U}[\mu - 3, \lambda + 3 + V(1)],$$

$$T^{33} \quad : \quad (\mathcal{U}[\mu, \chi])^{1 \times \alpha} \longmapsto \left(\mathcal{U}[\mu - 3, Y(\chi) + 3] \right)^{1 \times \alpha}. \tag{5.55}$$

Here, Y denotes the following mapping between sequences:

$$Y(\chi) := \left\{ \max \left[\chi_j, \max_{k < j} (\chi_k + 1) \right] \right\}_{j=0}^{\infty}. \tag{5.56}$$

Applying it to a constant sequence, gives:

$$Y(k) = \{k, k + 1, k + 1, \ldots\} = k + V(1).$$

The right inverse $A_0^{(-1)}$ built from (5.37) and (5.38) maps, due to (5.34) and (5.36), as follows:

$$\left(A_0^\square\right)^{(-1)} \quad : \quad \left(\mathcal{U}[\mu, \chi]\right)^{2\times\alpha} \longmapsto \left(\mathcal{U}[\mu+2, \chi]\right)^{2\times\alpha},$$

$$\left(A_0^{33}\right)^{(-1)} \quad : \quad \left(\mathcal{U}[\mu, \chi]\right)^{1\times\alpha} \longmapsto \left(\mathcal{U}[\mu+4, \max(\chi-2, 0)]\right)^{1\times\alpha}.$$

This implies for the particular blocks of the operator matrix (5.9) the mapping properties:

$$\left(A_0^\square\right)^{(-1)} T^\square \quad : \quad \left(\mathcal{U}[\mu, \chi]\right)^{2\times\alpha} \longmapsto \left(\mathcal{U}[\mu+1, \chi+2]\right)^{2\times\alpha} \tag{5.57}$$

$$\left(A_0^\square\right)^{(-1)} T_K \quad : \quad \left(\mathcal{U}[\mu, \chi]\right)^{1\times\alpha} \longmapsto \left(\mathcal{U}[\mu+1, \chi+1]\right)^{2\times\alpha} \tag{5.58}$$

$$\left(A_0^{33}\right)^{(-1)} T_K^\top \quad : \quad \left(\mathcal{U}[\mu, \chi]\right)^{2\times\alpha} \longmapsto \left(\mathcal{U}[\mu+3, \max(\chi-1, 0)]\right)^{1\times\alpha} \tag{5.59}$$

$$\left(A_0^{33}\right)^{(-1)} T^{33} \quad : \quad \left(\mathcal{U}[\mu, \chi]\right)^{1\times\alpha} \longmapsto \left(\mathcal{U}[\mu+1, Y(\chi)+1]\right)^{1\times\alpha} \tag{5.60}$$

Of course, lower powers of ρ^{-2} can occur in practice, if the nominator of the corresponding rational function is also a multiple of ρ^2. If we consider especially the subclasses

$$\mathcal{G}^*_{\mu,0} := \left\{\rho^2 p \ln\rho + q \;\middle|\; p \in \mathrm{Hom}^{(\mu-2)}, q \in \mathrm{Hom}^{(\mu)}\right\} \subset \mathcal{G}_{\mu,0},$$

we obtain sharper mapping properties, namely:

$$\tilde{\partial}_\alpha \quad : \quad \mathcal{G}^*_{\mu,0} \longmapsto \mathcal{G}_{\mu-1,0}, \tag{5.61}$$

$$\tilde{\Delta} \quad : \quad \mathcal{G}^*_{\mu,0} \longmapsto \mathcal{G}_{\mu-2,0},$$

$$\Delta_M - \tilde{\Delta} \quad : \quad \mathcal{G}^*_{\mu,0} \longmapsto \mathcal{U}[\mu-1, 1],$$

$$T^{33} \quad : \quad \mathcal{G}^*_{\mu,0} \longmapsto \mathcal{U}[\mu-3, 2+V(1)],$$

$$\left(A_0^{33}\right)^{(-1)} T^{33} \quad : \quad \mathcal{G}^*_{\mu,0} \longmapsto \mathcal{U}[\mu+1, V(1)]. \tag{5.62}$$

5.8 The Structure of the Series Terms

Formulas (5.57)–(5.62) confirm the result, already obtained by (2.24) and (2.28), that all entries of the matrix operator $A_0^{(-1)} T$ raise the degree of pseudohomogeneity at all positions at least by one step. Thus, the consecutive terms $N^{(j)}$ of the series (2.30) have a raising minimal degree, in other words, only finitely many summands contribute to each fixed degree of pseudohomogeneity.

The components $N_{ik}^{(j)}$ of the 3×3 matrices $N^{(j)}$ are series of the form (5.23). To which classes $\mathcal{U}[\mu, \chi]$ they belong, is specified in the following theorem. From there one can see, which maximal power of ρ^{-2} is assigned to each degree of pseudohomogeneity.

For preparation, we define sequences $\sigma^{(j)}$ by recursion:

$$\begin{aligned}
\sigma^{(0)} &= 0, \\
\sigma^{(1)} &= V(1) = \{0, 1, 1, 1, \ldots\}, \\
\sigma^{(j+1)} &= \max\left[V^2(2j-2), Y\left(\sigma^{(j)}\right) + 1\right], \quad \text{for } j \geq 1. \quad (5.63)
\end{aligned}$$

An explicit evaluation gives for $0 \leq j \leq 5$:

$$\sigma_l^{(j)} := \begin{cases} j + l - 1, & 0 \leq l \leq j \\ 2j - 1, & l \geq j \end{cases} \quad (5.64)$$

and for $j \geq 6$:

$$\sigma_l^{(j)} := \begin{cases} j + l - 1, & \text{for } l = 0, 1 \text{ and } j - 3 \leq l \leq j, \\ 2j - 4, & \text{for } 2 \leq l \leq j - 4, \\ 2j - 1, & \text{for } l \geq j. \end{cases} \quad (5.65)$$

Theorem 5.8.1. *For $j = 0, 1, 2, \ldots$ there holds:*

$$N_{\alpha\beta}^{(j)} \in \mathcal{U}[j, 2j + 1], \quad (5.66)$$

$$N_{\alpha 3}^{(j)} \in \mathcal{U}[j + 2, 2j - 1], \quad (5.67)$$

$$N_{3\alpha}^{(j)} \in \mathcal{U}[j + 2, 2j - 2], \quad (5.68)$$

$$N_{33}^{(j)} \in \mathcal{U}[j + 2, \sigma^{(j)}]. \quad (5.69)$$

Hereby $\mathcal{U}[j, \lambda] := \{0\}$ for $\lambda < 0$. The indices α and β take the values 1 and 2.

Proof. (By induction with respect to j.)
From (5.42), (5.44) and (5.45) follows the assertion for $j = 0$:

$$N_{\square}^{(0)} \in \mathcal{U}[0, 1], \quad N_{33}^{(0)} \in \mathcal{U}[2, 0], \quad N_{\alpha 3}^{(0)} = N_{3\alpha}^{(0)} = 0.$$

The stronger proposition

$$N_{33}^{(0)} \in \mathcal{G}_{2,0}^* \oplus \mathcal{U}[3, 0]$$

can be deduced from (5.45) as well. From the mapping properties (5.60) and (5.62) we get

$$N_{33}^{(1)} = \left(A_0^{33}\right)^{(-1)} T^{33} N_{33}^{(0)} \in \mathcal{U}\left[3, \sigma^{(1)}\right],$$

such that (5.69) is true for $j = 1$. It remains to verify the step from j to $j+1$, which is to perform for $j \geq 1$ if we consider (5.69), else for $j \geq 0$.

The mapping properties of the particular blocks of the operator matrix $A_0^{(-1)}T$ are described by (5.57)–(5.60). With (5.24), the corresponding sums can be evaluated. This is done below. Note that one of both summands in the following three expressions always vanishes for $j = 0$.

The upper left block of $N^{(j+1)}$ belongs to

$$\left(A_0^{\square}\right)^{(-1)} T^{\square} \left(\mathcal{U}[j, 2j+1]\right)^{2\times2} \oplus \left(A_0^{\square}\right)^{(-1)} T_K \left(\mathcal{U}[j+2, 2j-2]\right)^{1\times2}$$

$$= \left(\mathcal{U}[j+1, 2j+3]\right)^{2\times2} \oplus \left(\mathcal{U}[j+3, 2j-1]\right)^{2\times2}$$

$$= \left(\mathcal{U}[j+1, 2j+3]\right)^{2\times2}.$$

The upper right block of $N^{(j+1)}$ belongs to

$$\left(A_0^{\square}\right)^{(-1)} T^{\square} \left(\mathcal{U}[j+2, 2j-1]\right)^{2\times1} \oplus \left(A_0^{\square}\right)^{(-1)} T_K \, \mathcal{U}[j+2, \sigma^{(j)}]$$

$$= \left(\mathcal{U}[j+3, 2j+1]\right)^{2\times1} \oplus \left(\mathcal{U}[j+3, \sigma^{(j)}+1]\right)^{2\times1}$$

$$= \left(\mathcal{U}[j+3, 2j+1]\right)^{2\times1},$$

since all terms of the series $\sigma^{(j)}$ are $\leq 2j-1$. The lower left block of $N^{(j+1)}$ belongs to

$$\left(A_0^{33}\right)^{(-1)} T_K^{\top} \left(\mathcal{U}[j, 2j+1]\right)^{2\times2} \oplus \left(A_0^{33}\right)^{(-1)} T^{33} \left(\mathcal{U}[j+2, 2j-2]\right)^{1\times2}$$

$$= \left(\mathcal{U}[j+3, 2j]\right)^{1\times2} \oplus \left(\mathcal{U}[j+3, 2j-1+V(1)]\right)^{1\times2}$$

$$= \left(\mathcal{U}[j+3, 2j]\right)^{1\times2}.$$

The lower right entry of $N^{(j+1)}$ belongs (for $j \geq 1$) to

$$\left(A_0^{33}\right)^{(-1)} T_K^{\top} \left(\mathcal{U}[j+2, 2j-1]\right)^{2\times1} \oplus \left(A_0^{33}\right)^{(-1)} T^{33} \, \mathcal{U}[j+2, \sigma^{(j)}]$$

$$= \mathcal{U}[j+5, 2j-2] \oplus \mathcal{U}[j+3, Y(\sigma^{(j)})+1]$$

$$= \mathcal{U}[j+3, \sigma^{(j+1)}].$$

Here we have used the recursion formula (5.63). The induction step from j to $j+1$ is complete. ∎

Corollary 5.8.1. *The fundamental solution $G(x, y)$ possesses a convergent series expansion within a sufficiently small neighbourhood of the point y. The particular terms of the series are pseudohomogeneous functions whose degree raises step-by-step. The components of G belong to the classes:*

$$G_{\alpha\beta} \in \mathcal{U}\left[0, \{2j+1\}_{j=0}^{\infty}\right], \tag{5.70}$$

$$G_{\alpha3} \in \mathcal{U}\left[3, \{2j+1\}_{j=0}^{\infty}\right], \tag{5.71}$$

$$G_{3\alpha} \in \mathcal{U}\left[3, \{2j\}_{j=0}^{\infty}\right], \tag{5.72}$$

$$G_{33} \in \mathcal{U}[2, \sigma], \tag{5.73}$$

where σ denotes the sequence $\sigma = \{0, 0, 1, 2, 3, 4, 5, 6, 8, 10, 12, 14, \ldots\}$ (after the 6 it contains all even numbers).

Proof. In sections 2.3 and 2.7 we have already shown that the fundamental solution possesses a local convergent series expansion and that its pseudohomogeneous terms belong to the same classes like the corresponding terms of the Levi function.

Formulas (5.70)–(5.72) are obvious from (5.66)–(5.68). For the proof of (5.73) we need the relation

$$\sigma_l = \max_{0 \le j \le l} \sigma_{l-j}^{(j)}$$

for the elements of the sequence σ. It can easily be verified if the sequences $\sigma^{(j)}$ are written one below the other and compared.   ∎

5.9 Pseudohomogeneous Expansion of the Fundamental Solution

Now we collect the results from the previous sections. A fundamental solution can be written as 3×3-matrix

$$G(x,y) = \|G_{ki}(x,y)\|_{i,k=1}^{3}, \tag{5.74}$$

which depends on the four variables x_1, x_2, y_1, y_2. As can be seen from (5.70)–(5.73), the individual matrix components admit series expansions with pseudohomogeneous terms as follows (α and β take the values 1 and 2):

$$G_{\alpha\beta} = \left(p_{\alpha\beta}^{(0)} + p_{\alpha\beta}^{(1)} + \cdots\right)\ln\rho + \frac{q_{\alpha\beta}^{(2)}}{\rho^2} + \frac{q_{\alpha\beta}^{(7)}}{\rho^6} + \frac{q_{\alpha\beta}^{(12)}}{\rho^{10}} + \frac{q_{\alpha\beta}^{(17)}}{\rho^{14}} + \cdots$$

$$G_{\alpha 3} = \left(p_{\alpha 3}^{(3)} + p_{\alpha 3}^{(4)} + \cdots\right)\ln\rho + \frac{q_{\alpha 3}^{(5)}}{\rho^2} + \frac{q_{\alpha 3}^{(10)}}{\rho^6} + \frac{q_{\alpha 3}^{(15)}}{\rho^{10}} + \frac{q_{\alpha 3}^{(20)}}{\rho^{14}} + \cdots$$

$$G_{3\alpha} = \left(p_{3\alpha}^{(3)} + p_{3\alpha}^{(4)} + \cdots\right)\ln\rho + q_{3\alpha}^{(3)} + \frac{q_{3\alpha}^{(8)}}{\rho^4} + \frac{q_{3\alpha}^{(13)}}{\rho^8} + \frac{q_{3\alpha}^{(18)}}{\rho^{12}} + \cdots$$

$$G_{33} = \left(p_{33}^{(2)} + p_{33}^{(3)} + \cdots\right)\ln\rho + q_{33}^{(2)} + q_{33}^{(3)} + \frac{q_{33}^{(6)}}{\rho^2} + \frac{q_{33}^{(9)}}{\rho^4} + \cdots + \frac{q_{33}^{(21)}}{\rho^{12}}$$
$$+ \frac{q_{33}^{(26)}}{\rho^{16}} + \frac{q_{33}^{(31)}}{\rho^{20}} + \frac{q_{33}^{(36)}}{\rho^{24}} + \cdots$$

Here $p_{ik}^{(j)}$ and $q_{ik}^{(j)}$ stand for homogeneous polynomials in $\vartheta = x - y$ of degree j. The coefficients of these polynomials depend on y and can be calculated explicitly. The variable ρ, defined by (5.15), coincides (up to the factor $\sqrt[4]{a}$) for $x \approx y$ in first approximation with the distance of the mid-surface points $\mathbf{r}(x)$ and $\mathbf{r}(y)$.

Note a special feature in the above representation for G_{33}. The powers of ρ^{-1} raise first in each step by 2, and after the 12, in each step by 4. Then the influence of the upper left block, by virtue of the coupling, is in force.

5.10 Form of the Remainder

Now we investigate the remainder for the special case that the Levi function consists of the upper left block (5.44) and the entry (5.46) on the position L_{33}. The remaining four positions are set to zero. The remainder $R(x, y)$ is defined by (2.8) and has the components

$$R^j_{.k} = \delta^j_k \, \delta(x - y) - A^{jl}(x, \partial) L_{lk}(x, y). \tag{5.75}$$

Here and in the sequel, the row index of the remainder is an upper one and the column index is a lower one. This is for reasons of compatibility with the tensor notation and the summation convention.

5.10.1 The Upper Left Block

If we write all differentiations in (4.49) according to the chain rule, this operator gets the form

$$A^\square = M_2(x)\partial^2 + M_1(x)\partial^1 + M_0(x).$$

Here M_j stands for certain multiplication operators and ∂^j symbolizes differentiations of order j. The application to the 2×2 block (5.44) (denoted with $L^\square$ in what follows) gives:

$$
\begin{aligned}
R^\square &:= \delta(x - y)I_{2\times2} - A^\square L^\square = \\
&= [M_2(y) - M_2(x)]\,\partial^2 L^\square - M_1(x)\partial^1 L^\square - M_0(x)L^\square.
\end{aligned}
$$

Each entry of $L^\square$ belongs to $\mathcal{G}_{0,1}$. Then (5.48) implies that the entries of $R^\square$ belong to $\mathcal{U}[-1, 3]$. The third power of ρ^{-2} occurs indeed and can not be cancelled, except for an isothermal parametrization of the mid-surface. In this case, however, both the terms with ρ^{-6} and ρ^{-4} vanish, $\underline{M}$ becomes the identity and the application of the operator (4.57) to $L^\square$ gives

$$
A^\square L^\square = \frac{1}{2\pi}\begin{pmatrix} \partial_1 & -\partial_2 \\ \partial_2 & \partial_1 \end{pmatrix}\left[\frac{\sqrt{a(y)}}{\sqrt{a(x)}\,\rho^2}\begin{pmatrix} \vartheta^1 & \vartheta^2 \\ -\vartheta^2 & \vartheta^1 \end{pmatrix}\right] + 2D\check{v}\mathcal{K}L^\square =
$$

$$
= \delta(x - y)I - \frac{\sqrt{a(y)}}{4\pi\sqrt{a(x)}^3}\begin{pmatrix} a_{,1}(x) & -a_{,2}(x) \\ a_{,2}(x) & a_{,1}(x) \end{pmatrix}\frac{1}{\rho^2}\begin{pmatrix} \vartheta^1 & \vartheta^2 \\ -\vartheta^2 & \vartheta^1 \end{pmatrix} + 2D\check{v}\mathcal{K}L^\square.
$$

As can be seen, higher powers of ρ^{-2} don't occur for an isothermal parametrization.

5.10.2 The Lower Right Entry

The function (5.46) contains pseudohomogeneous terms of degree 2 and 3:

$$L_{33} = L_{33}^{(2)} + L_{33}^{(3)}, \quad \text{with } L_{33}^{(2)} \in \mathcal{G}_{2,0}^*, \ L_{33}^{(3)} \in \mathcal{G}_{3,0}.$$

The operator (4.52), which is applied to L_{33}, has the symbolic representation

$$A^{33} = M_4(x)\partial^4 + M_3(x)\partial^3 + M_2(x)\partial^2 + M_1(x)\partial^1 + M_0(x).$$

The function (5.46) is constructed just in such a way that $M_4(y)\partial^4 L_{33}^{(2)} = \delta(x - y)$ and the terms of degree -1 cancel out:

$$(x - y)M_4'(y)\partial^4 L_{33}^{(2)} + M_4(y)\partial^4 L_{33}^{(3)} + M_3(y)\partial^3 L_{33}^{(2)} = 0.$$

One of the pseudohomogeneous terms of degree zero is

$$(x - y)M_4'(y)\partial^4 L_{33}^{(3)} \in \mathcal{G}_{0,4}.$$

It produces the highest powers of ρ^{-2} in the remainder, namely ρ^{-8}. Therefore $R_{\cdot3}^3$ belongs to $\mathcal{U}[0, 4]$.

Only the powers ρ^{-2} occur for an isothermal parametrization as will be shown below. In this case, (5.46) coincides with (5.45) and Christoffel symbols have the special form (4.56). From

$$\Delta\left[\rho^2 \left(\ln\rho - 1\right)\right] = 4\ln\rho,$$

$$\Delta\left[\vartheta^\alpha \rho^2 \left(\ln\rho - \frac{3}{4}\right)\right] = 8\vartheta^\alpha \ln\rho$$

it follows that

$$\Delta L_{33} = -\frac{a(y)}{2\pi B}\left[1 + \vartheta^\alpha \Gamma_{\alpha\beta}^\beta(y)\right]\ln\rho,$$

$$A^{33}L_{33} = \frac{1}{2\pi}\Delta\left\{\sqrt{\frac{a(y)}{a(x)}}\left[1 + \vartheta^\alpha \Gamma_{\alpha\beta}^\beta(y)\right]\ln\rho\right\} -$$

$$- 2B\tilde{\nu}\left(\mathcal{K}\Delta L_{33} + \mathcal{K}_{,1}L_{33,1} + \mathcal{K}_{,2}L_{33,2}\right) + 4D\sqrt{a(x)}\left(\tilde{\nu}\mathcal{K} - \mathcal{H}^2\right)L_{33}$$

(cmp. (4.58)). Taking into account the mapping property (5.49), we conclude $R_{\cdot3}^3 \in \mathcal{U}[0, 1]$.

5.10.3 The Upper Right Block

Recall that $L_{33}^{(3)} \in \mathcal{G}_{3,0}$ and, for an isothermal parametrization, $L_{33}^{(3)} \in \mathcal{G}_{3,0}^*$. The operator (4.50) contains only first derivatives. The mapping properties (5.48) and (5.61) imply $R_{\cdot3}^\alpha \in \mathcal{U}[1, \{0, 1, 1, 1, \ldots\}]$ for the general case and $R_{\cdot3}^\alpha \in \mathcal{U}[1, 0]$ for isothermal parametrization.

5.10.4 The Lower Left Block

Application of (4.51) to $L^{\square}$ gives functions from $\mathcal{U}[-1, 2]$. The powers ρ^{-4} occure also for an isothermal parametrization as can be seen from the example considered in chapter 7 (see formula (7.4)).

6. The System of Integral Equations and its Numerical Solution

The Dirichlet problem for the DV model, in which case the slope and three displacement components on the lateral boundary are given, will be transformed into a system of boundary and domain integral equations by means of the *indirect method*. Mapping and solvability properties of this system are studied in appropriate Sobolev spaces. The corner singularities are taken into account.

A Galerkin method for numerical solution of the system is studied on quasi-uniform and on graded meshes. For both versions, the reachable order of convergence is determined. In order to reduce the computational time for the discretization, we give an estimate for the admissible quadrature errors such that the order of convergence is not disturbed.

6.1 Boundary Geometry and the Integral Theorem of Gauss

The pre-image Ω of the mapping (4.2) is assumed to be an open, simple connected, bounded subset of $\mathbb{R}^2$. Further we assume that the boundary Γ is a regular curvilinear polygonal as it was introduced in Definition 3.7.1. The algorithm and the analytical investigations become significantly simpler for smooth boundaries, but this would restrict the applicability of the method strongly. Firstly, the boundaries of real shells often have corners; secondly, the considered cut-out may origin from a decomposition of a complex thin structure. The later case is of special interest since domain decomposition methods and coupling methods are very popular for the numerical investigation of complex structures.

In the sequel we consider besides Γ also its image $\mathbf{r}(\Gamma)$, i. e., the boundary of the shell mid-surface. It is called the *mid-surface boundary*. Let $[\Gamma_1(s), \Gamma_2(s)]$ be a parametrization of Γ in positive orientation. Due to the assumptions, the derivative

$$\frac{d\mathbf{r}\,(\Gamma_1(s), \Gamma_2(s))}{ds} = \mathbf{a}_1 \Gamma_1'(s) + \mathbf{a}_2 \Gamma_2'(s) \tag{6.1}$$

is defined almost everywhere and represents a tangential vector for the mid-surface boundary with the length

$$\gamma(s) := \left[a_{11} \left(\Gamma_1' \right)^2 + 2a_{12} \Gamma_1' \Gamma_2' + a_{22} \left(\Gamma_2' \right)^2 \right]^{\frac{1}{2}}. \qquad (6.2)$$

Thus

$$d\tilde{\Gamma} := \gamma(s)\, ds$$

is the arc-length measure on the mid-surface boundary.

The tangential unit vector $\mathbf{t}$ is obtained by multiplication of the vector (6.1) with γ^{-1}. It has the contravariant components

$$\begin{pmatrix} t^1 \\ t^2 \end{pmatrix} = \begin{pmatrix} \dot{\Gamma}_1 \\ \dot{\Gamma}_2 \end{pmatrix} := \frac{1}{\gamma(s)} \begin{pmatrix} \Gamma_1'(s) \\ \Gamma_2'(s) \end{pmatrix}$$

and the covariant components $t_\alpha = a_{\alpha\beta} t^\beta$.

The in-plane, outer unit normal vector to the mid-surface boundary is $\mathbf{n} := \mathbf{t} \times \mathbf{a}_3$ (see Fig. 6.1). Its components are

$$\begin{pmatrix} n_1 \\ n_2 \end{pmatrix} = \sqrt{a} \begin{pmatrix} \dot{\Gamma}_2 \\ -\dot{\Gamma}_1 \end{pmatrix}, \qquad n^\alpha = a^{\alpha\beta} n_\beta.$$

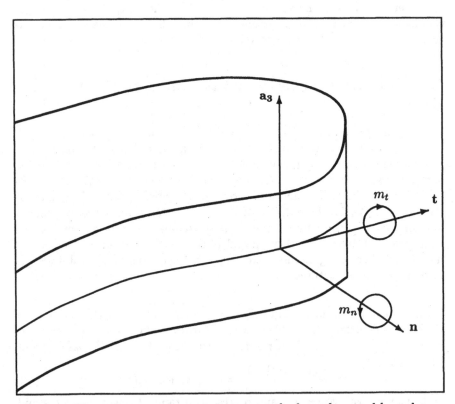

Fig. 6.1. The orthonormal vector system on the boundary and boundary moments

The Cartesian components of the outer unit normale vector to Γ in the parameter plane are:

$$\frac{ds}{d\Gamma} \begin{pmatrix} \Gamma_2'(s) \\ -\Gamma_1'(s) \end{pmatrix}.$$

(6.3)

Lemma 6.1.1. *For an isothermal parametrization, the interiour angles of the polygon Γ coincide with the interiour angles of the mid-surface boundary.*

Proof. For an isothermal parametrization holds

$$\begin{pmatrix} t^1 \\ t^2 \end{pmatrix} = \frac{1}{\sqrt[4]{a}\sqrt{(\Gamma_1')^2 + (\Gamma_2')^2}} \begin{pmatrix} \Gamma_1' \\ \Gamma_2' \end{pmatrix} =: \frac{1}{\sqrt[4]{a}} \mathbf{t}_\Gamma,$$

where $\mathbf{t}_\Gamma$ denotes the tangential unit vector to Γ in the parameter plane. If we take the covariant components of $\mathbf{t}$ and form the vector $(t_1, t_2)^\top \in \mathbb{R}^2$ from them, we obtain the identity $(t_1, t_2)^\top = \sqrt[4]{a}\,\mathbf{t}_\Gamma$.

For corner points, we denote the right-side and left-side limits with an upper "+" or "-", respectively. For the scalar products of the tangential vectors holds obviously

$$(\mathbf{t}^+, \mathbf{t}^-) = (\mathbf{t}_\Gamma^+, \mathbf{t}_\Gamma^-).$$

This means, the cosine of an interiour angle on Γ coincides with the cosine of the corresponding angle on the mid-surface boundary. The angles itself are equal as well. This can be seen e. g. from the coicidence of the scalar product $(\mathbf{t}^+, \mathbf{n}^-)$ with the corresponding scalar product in the parameter plane. ∎

Let $\mathbf{w}(x) = w^\alpha(x)\mathbf{a}_\alpha(x)$ be an arbitrary tangential vector field on the mid-surface with integrable first derivatives. Using the notation

$$d\tilde{\Omega} := \sqrt{a(x)}\, dx_1\, dx_2$$

for the area measure on the mid-surfce, we can write the integral theorem of Gauss in the form (cmp. [9, sect. 1.5.2]):

$$\int_\Omega w^\alpha|_\alpha\, d\tilde{\Omega} = \int_\Gamma w^\alpha n_\alpha d\tilde{\Gamma}.$$

(6.4)

This can be verified by use of (4.4), (4.5) and (6.3):

$$\int_\Omega w^\alpha|_\alpha\, d\tilde{\Omega} = \int_\Omega (w^\alpha_{,\alpha} + \Gamma^\alpha_{\alpha\beta}w^\beta)\sqrt{a}\, dx = \int_\Omega (\sqrt{a}\, w^\alpha)_{,\alpha}\, dx$$

$$= \int_\Gamma \sqrt{a}\,(w^1\Gamma_2' - w^2\Gamma_1')\, ds = \int_\Gamma w^\alpha n_\alpha d\tilde{\Gamma}.$$

Now, let us determine the directional derivatives of a scalar function $w(x)$ along $\mathbf{t}$ and $\mathbf{n}$. The scalar product with the gradient $w_{,\alpha}\mathbf{a}^\alpha$ gives

$$w_{,t} := \frac{\partial w}{\partial t} = w_{,\alpha}t^\alpha, \qquad w_{,n} := \frac{\partial w}{\partial n} = w_{,\alpha}n^\alpha$$

(6.5)

for the tangential and normal derivative, respectively. We can re-arrange these equations as follows:

$$w_{,\alpha} = n_\alpha w_{,n} + t_\alpha w_{,t}.$$ (6.6)

If w is a continuously differentiable function, then the left side of (6.6) is continuous even at corner points, both summands on the right side, however, are discontinuous in general.

6.2 Boundary Data and Green's Formula

If we multiply the equations of equilibrium (4.11), (4.12) with a triplet of test functions $(u_1, u_2, u_3) \in \left[C^\infty(\overline{\Omega}) \right]^3$, integrate over Ω and apply the integral theorem of Gauss, we obtain:

$$\int_\Omega \left[n^{(\alpha\beta)} \Big|_\beta u_\alpha + \left(q^\alpha \big|_\alpha + b_{\alpha\beta} n^{(\alpha\beta)} \right) u_3 \right] d\tilde{\Omega}$$

$$= \int_\Omega \left[\left(n^{(\alpha\beta)} u_\alpha + q^\beta u_3 - m^{(\alpha\beta)} u_{3,\alpha} \right) \Big|_\beta - \right.$$
$$\left. - n^{(\alpha\beta)} \left(u_{\alpha|\beta} - b_{\alpha\beta} u_3 \right) + m^{(\alpha\beta)} u_{3|\alpha\beta} \right] d\tilde{\Omega}$$

$$= \int_\Gamma \left(n^{(\alpha\beta)} u_\alpha + q^\beta u_3 - m^{(\alpha\beta)} u_{3,\alpha} \right) n_\beta d\tilde{\Gamma} - E(u,v) \quad (6.7)$$

$$= - \int_\Omega p^j u_j d\tilde{\Omega}.$$

Here, $E(u,v)$ is the symmetric energy bilinear form (4.74).

In order to bring the boundary integral from (6.7) into another form, we introduce operators which map onto the Dirichlet and Neumann data. These are not uniquely determined, as well-known. We will give two different variants, at first the pair $\left(\widehat{S}, \widehat{T} \right)$ which maps onto the so called "physical boundary data" and secondly the pair (S, T) which maps onto an other set of boundary data, well-adapted to the Dirichlet problem and more convenient for analytical and numerical purposes in this case.

The set of Dirichlet data

$$\widehat{S}(u) := \begin{pmatrix} u_n \\ u_t \\ u_3 \\ u_{3,n} \end{pmatrix} := \begin{pmatrix} u_\alpha n^\alpha \\ u_\alpha t^\alpha \\ u_3 \\ u_{3,\alpha} n^\alpha \end{pmatrix}$$

contains the components of the boundary displacement with respect to the local base $(\mathbf{n}, \mathbf{t}, \mathbf{a}_3)$ and the boundary slope $u_{3,n}$. The corresponding set of Neumann data

$$\widehat{T}(v) = (\tau_n, \tau_t, \tau_3, -m_t)^\top$$ (6.8)

contains the real boundary forces in normal and tangential direction

$$T_n := n^{(\alpha\beta)} n_\alpha n_\beta \quad \text{und} \quad T_t := n^{(\alpha\beta)} n_\alpha t_\beta;$$

further the "equivalent shear force" (or Kirchhoff shear)

$$T_3 := q^\alpha n_\alpha - \gamma^{-1} \frac{d\, m_n}{ds} \tag{6.9}$$

along the shell normale a_3 and the tangential boundary moment

$$m_t := m^{(\alpha\beta)} n_\alpha n_\beta.$$

The moment

$$m_n := -m^{(\alpha\beta)} n_\alpha t_\beta$$

along the axis n (see fig. 2) can be pre-described independently only for models without Kirchhoff-Love hypothesis.

If the mid-surface boundary contains corners, the components of n and t are discontinuous at these points. The same holds true for m_n. The derivative of m_n in (6.9) is to be understood in distributional sense. The jumps of m_n generate point functionals, the so called "corner forces". Therefore it is necessary to require continuity of the test function u_3 at the corner points. Partial integration can be applied then:

$$\int_\Gamma \frac{dm_n}{ds} u_3 ds = -\int_\Gamma m_n \frac{du_3}{ds} ds = -\int_\Gamma m_n u_{3,t} d\tilde{\Gamma}. \tag{6.10}$$

This formula is used in the following re-formulation of the boundary integral from (6.7):

$$\int_\Gamma \left(n^{(\alpha\beta)} u_\alpha + q^\beta u_3 - m^{(\alpha\beta)} u_{3,\alpha} \right) n_\beta d\tilde{\Gamma}$$

$$= \int_\Gamma \left[n^{(\alpha\beta)} n_\beta \left(n_\alpha u_n + t_\alpha u_t \right) + \left(T_3 + \gamma^{-1} \frac{dm_n}{ds} \right) u_3 - \right.$$

$$\left. - m^{(\alpha\beta)} \left(n_\alpha u_{3,n} + t_\alpha u_{3,t} \right) n_\beta \right] d\tilde{\Gamma}$$

$$= \int_\Gamma \left[T_n u_n + T_t u_t + T_3 u_3 - m_t u_{3,n} \right] d\tilde{\Gamma}.$$

A Green's formula of the first kind for the system (4.14) takes the form:

$$\int_\Omega u_j \left(A^{jk} v_k \right) dx = -E(u,v) + \int_\Gamma \left[\hat{T}(v) \right]^T \hat{S}(u) d\tilde{\Gamma}. \tag{6.11}$$

This relation is called the *principle of virtual energy* in mechanics, since both sides of the equation can be interpreted as the energy generated by a virtual displacement field $u = (u_1, u_2, u_3)$.

The corner forces don't occur if the expression $q^\beta n_\beta u_3$ in (6.7) is partial integrated in tangential direction. Note that the shear forces q^α depend only on v_3. Let Q_0 denote the functional which maps the displacement component v_3 to the mean value of $q^\beta n_\beta$ via

$$Q_0(v_3) := \frac{\int_\Gamma q^\beta n_\beta d\tilde{\Gamma}}{\int_\Gamma d\tilde{\Gamma}}. \tag{6.12}$$

Further, denote by $Q(s)$ the periodic function defined by

$$\frac{dQ}{d\tilde{\Gamma}} = \gamma^{-1}(s) \frac{dQ(s)}{ds} = q^\beta n_\beta - Q_0. \tag{6.13}$$

Then holds

$$\int_\Gamma q^\beta n_\beta u_3 d\tilde{\Gamma} = Q_0(v_3) < u_3, 1 > - \int_\Gamma Q u_{3,t} d\tilde{\Gamma},$$

$$< u_3, 1 > := \int_\Gamma u_3 d\tilde{\Gamma}.$$

This leads to the Green's formula of the first kind

$$\int_\Omega u_j \left(A^{jk} v_k\right) dx = -E(u,v) + \int_\Gamma [T(v)]^\mathsf{T} S(u) \, d\tilde{\Gamma} + Q_0(v_3) < u_3, 1 > \tag{6.14}$$

with boundary data

$$S(u) := \mathrm{tr}_x S_x u(x) = \left(u_1, \ u_2, \ u_{3,1}, \ u_{3,2}\right)^\mathsf{T}, \tag{6.15}$$

$$T(v) := \left(n^{(1\beta)} n_\beta, \ n^{(2\beta)} n_\beta, \ -Q - m^{(1\beta)} n_\beta, \ -Q - m^{(2\beta)} n_\beta\right)^\mathsf{T}. \tag{6.16}$$

Here S_x denotes the 4×3-matrix

$$S_x := \begin{pmatrix} 1 & 0 & 0 \\ 0 & 1 & 0 \\ 0 & 0 & \frac{\partial}{\partial x_1} \\ 0 & 0 & \frac{\partial}{\partial x_2} \end{pmatrix}. \tag{6.17}$$

It is necessary to distinguish between $S(u)$ and $S_x u$. The later does not contain the trace operator!

The function Q defined by (6.13) contains still a free integration constant. According to this, each quadrupel of Dirichlet data (6.15) has to fulfill one combatibilty condition, since

$$\int_\Gamma u_{3,t} d\tilde{\Gamma} = \int_\Gamma u_{3,\alpha} t^\alpha d\tilde{\Gamma} = \int_\Gamma (u_{3,1} \Gamma_1' + u_{3,2} \Gamma_2') \, ds = 0. \tag{6.18}$$

By virtue of the symmetry of the energy bilinear form $E(u,v)$, we can deduce from (6.11) and (6.14) the following two variants for Green's formula of the second kind:

$$\int_{\Omega} \left(u_j A^{jk} v_k - v_j A^{jk} u_k \right) dx = \int_{\Gamma} \left\{ \left[\widehat{T}(v) \right]^{\mathsf{T}} \widehat{S}(u) - \left[\widehat{T}(u) \right]^{\mathsf{T}} \widehat{S}(v) \right\} d\tilde{\Gamma}$$

$$= \int_{\Gamma} \left\{ \left[T(v) \right]^{\mathsf{T}} S(u) - \left[T(u) \right]^{\mathsf{T}} S(v) \right\} d\tilde{\Gamma} +$$

$$+ Q_0(v_3) < u_3, 1 > - Q_0(u_3) < v_3, 1 > .$$

6.3 The Integral Equation System for the Dirichlet Problem

6.3.1 The Dirichlet Problem

The system (4.48) of partial differential equations gives, together with the Dirichlet data (6.15), the boundary value problem:

$$\left. \begin{array}{ll} A(x,\partial)v(x) = -p(x), & x \in \Omega, \\ \mathbf{tr}_x S_x v(x) = \phi(x), & x \in \Gamma. \end{array} \right\} \tag{6.19}$$

Here $v(x)$ denotes the vector with components v_1, v_2, v_3. The right-hand side of (4.48) is denoted by $-p(x)$ and

$$\phi(x) := [\phi_1(x), \phi_2(x), \phi_3(x), \phi_4(x)]^{\mathsf{T}}$$

are the Dirichlet data. They have to fulfill the side condition

$$\int_{\Gamma} [\phi_3(x) \, dx_1 + \phi_4(x) \, dx_2] = 0, \tag{6.20}$$

as can be seen from (6.18).

6.3.2 Transformation into a System of Integral Equations

We will use (slightly modified) the indirect method as described in section 3.1. The solution $v(x)$ of the boundary value problem (6.19) is searched for in the form

$$v(x) = \int_{\Omega} L^{\Omega}(x,y)w(y)dy + \int_{\Gamma} L^{\Gamma}(x,y)f(y)d\Gamma_y, \tag{6.21}$$

i. e., as the sum of a domain integral and a simple layer potential. The vector functions

$$w(y) = \begin{pmatrix} w^1(y) \\ w^2(y) \\ w^3(y) \end{pmatrix}, \qquad f(y) = \begin{pmatrix} f^1(y) \\ f^2(y) \\ f^3(y) \\ f^4(y) \end{pmatrix},$$

which are to be determined, have no physical meaning in this method.

For L^Ω we use the Levi function from section 5.6:

$$L^\Omega = \begin{pmatrix} L^\square & 0 \\ 0 & L_{33} \end{pmatrix} \in \Psi\overline{\Omega} \begin{pmatrix} 0 & 0 & / \\ 0 & 0 & / \\ / & / & 2 \end{pmatrix}. \tag{6.22}$$

The 2×2-matrix $L^\square = L^\square(x,y)$ is given by (5.44). (The term $(1+\nu)/2$ in the square bracket will be neglected.) On the lower right position $L_{33} = L_{33}(x,y)$ stands the function (5.46) (without polynomial terms).

In (6.22) occurs a new notation which is used in the sequel for matrices of functions from $\Psi\overline{\Omega}(\mu)$. The individual integers show the minimal degree of pseudohomogeneity of the corresponding function on this position. The sign "/" is used for an entry identical to zero.

The other kernel in (6.21) is the 3×4-matrix

$$L^\Gamma(x,y) := \left\{ S_y^\diamond \left[L^\Omega(x,y) \right]^\top \right\}^\top = \tag{6.23}$$

$$= \begin{pmatrix} L^\square & 0 & 0 \\ 0 & \partial_1^\diamond L_{33} & \partial_2^\diamond L_{33} \end{pmatrix} \in \Psi\overline{\Omega} \begin{pmatrix} 0 & 0 & / & / \\ 0 & 0 & / & / \\ / & / & 1 & 1 \end{pmatrix}$$

with

$$S_y^\diamond := \begin{pmatrix} 1 & 0 & 0 \\ 0 & 1 & 0 \\ 0 & 0 & \partial_1^\diamond \\ 0 & 0 & \partial_2^\diamond \end{pmatrix}, \qquad \partial_\alpha^\diamond := \sqrt{a(y)}\, \frac{\partial}{\partial y_\alpha} \frac{1}{\sqrt{a(y)}}.$$

The multiplication with $\sqrt{a(y)}$ and its inverse is involved for convenience. Then the formulas for L^Γ and the integral kernels derived from it become shorter. The more natural way would be to use the area measure $d\tilde{\Omega}_y = \sqrt{a(y)}\,dy$ instead of dy in (6.21). In order to get shorter formulas, the function $\sqrt{a(y)}$ it is already contained in L^Ω.

Application of the differential operator $A(x,\partial)$ and the operator $\mathrm{tr}_x S_x$ to the ansatz (6.21) leads to the integral equation system:

$$w(x) - \int_\Omega R^\Omega(x,y)w(y)dy - \int_\Gamma R^\Gamma(x,y)f(y)d\Gamma_y = -p(x), \quad \forall x \in \Omega,$$

$$\int_\Omega V^\Omega(x,y)w(y)dy + \int_\Gamma V^\Gamma(x,y)f(y)d\Gamma_y = \phi(x), \quad \forall x \in \Gamma. \tag{6.24}$$

6.3.3 The Kernel Functions

The form of the remainder $R^\Omega(x,y)$ was already determined in section 5.10:

$$R^\Omega(x,y) \in \Psi\overline{\Omega} \begin{pmatrix} -1 & -1 & 1 \\ -1 & -1 & 1 \\ -1 & -1 & 0 \end{pmatrix}.$$

This gives

$$R^\Gamma(x,y) = \left\{ S_y^\circ \left[R^\Omega(x,y) \right]^\top \right\}^\top = \tag{6.25}$$

$$= -A(x,\partial)L^\Gamma(x,y) \in \Psi\overline{\Omega} \begin{pmatrix} -1 & -1 & 0 & 0 \\ -1 & -1 & 0 & 0 \\ -1 & -1 & -1 & -1 \end{pmatrix}.$$

Now it is obvious, why it is meaningful to use a Levi function of non-uniform degree and to add the pseudohomogeneous term of degree 3 to the lower right position of L^Ω. If one uses only the leading singularity here, then $R_{33}^\Gamma(x,y)$ and $R_{34}^\Gamma(x,y)$ would contain terms of degree -2, which involves hypersingular integrals already for the solution of the Dirichlet problem. (For the Neumann problem, hypersingular integrals can not be avoided in any case.)

The kernel functions of the boundary integral equation (6.24) are:

$$V^\Omega(x,y) = S_x L^\Omega(x,y) = \begin{pmatrix} L^\square & 0 \\ 0 & L_{33,1} \\ 0 & L_{33,2} \end{pmatrix} \in \Psi\overline{\Omega} \begin{pmatrix} 0 & 0 & / \\ 0 & 0 & / \\ / & / & 1 \\ / & / & 1 \end{pmatrix}, \tag{6.26}$$

$$V^\Gamma(x,y) = S_x L^\Gamma(x,y) = \begin{pmatrix} L^\square & 0 \\ 0 & L_\square \end{pmatrix} \in \Psi\overline{\Omega} \begin{pmatrix} 0 & 0 & / & / \\ 0 & 0 & / & / \\ / & / & 0 & 0 \\ / & / & 0 & 0 \end{pmatrix}, \tag{6.27}$$

with

$$L_\square := \begin{pmatrix} \partial_1 \partial_1^\circ & \partial_1 \partial_2^\circ \\ \partial_2 \partial_1^\circ & \partial_2 \partial_2^\circ \end{pmatrix} L_{33}(x,y) =$$

$$= \frac{\sqrt{a(y)}}{4\pi B} \underline{M}^{-1}(y) \left[\ln \rho + \frac{1}{\rho^2} \begin{pmatrix} (\vartheta^1)^2 & \vartheta^1\vartheta^2 \\ \vartheta^1\vartheta^2 & (\vartheta^2)^2 \end{pmatrix} \underline{M}^{-1}(y) \right] +$$

$$+ \text{ terms from } \Psi\overline{\Omega} \begin{pmatrix} 1 & 1 \\ 1 & 1 \end{pmatrix}.$$

The verification of the last relation requires some calculation and uses formulas from differential geometry, collected in chapter 4.

6.3.4 The Enhanced System of Integral Equations

The boundary integral operator V^Γ is not invertible. From

$$f^0(y) := \frac{1}{\sqrt{a(y)}} \begin{pmatrix} 0 \\ 0 \\ \frac{dy_1}{d\Gamma} \\ \frac{dy_2}{d\Gamma} \end{pmatrix} \in \ker V^\Gamma \tag{6.28}$$

and (6.20) follows that the zero space and the cokernel of V^Γ is non-trivial, but at least of dimension one. This is a well-known phenomenon which appears as well for the biharmonic equation, if the Dirichlet data are pre-designed in form of its partial derivatives. (The close relationship of the operator L^{33} to the biharmonic operator is obvious from (4.52).)

According to the indirect boundary integral method for elliptic equations of higher order (with known fundamental solution) described in [38], the non-invertibility can be overcome by the introduction of additional unknowns and additional equations. Later, this method was worked out and analyzed in [64] for the special case of higher powers of the Laplacian.

The papers [65] and [66] are devoted to the biharmonic equation and some related systems. The Dirichlet data is assumed to be given there in form of both partial derivatives (instead of the trace of the function and its normal derivative). The simple layer potential is used in this papers with 3 additional equations. It appears however, that the zero space and the cokernel is one-dimensional, such that one additional equation suffices.

According to this knowledge, we modify the boundary integral system (6.24) as follows: one additional (scalar) unknown f_0 is introduced and the vector $e_0 f_0$ with

$$e_0 := (0, 0, -x_2, x_1)^\top \tag{6.29}$$

is added to the left side of (6.24). The system is enhanced by the additional equation

$$E_0 f := \int_\Gamma \left[\frac{dy_1}{d\Gamma} f^3(y) + \frac{dy_2}{d\Gamma} f^4(y) \right] d\Gamma_y = \phi_0. \tag{6.30}$$

One can choose $\phi_0 = 0$. Other values lead to different solution vectors f, the displacement field, however, remains unchanged. The described modification leads to one additional row and one additional column in the disretization matrix. The solution of the obtained linear system must generate the value $f_0 = 0$, at least approximately. This criterion can be applied among others to check the numerical solution.

The complete enhanced system takes the form

$$\begin{pmatrix} I - R^\Omega & -R^\Gamma & 0 \\ V^\Omega & V^\Gamma & e_0 \\ 0 & E_0 & 0 \end{pmatrix} \begin{pmatrix} w \\ f \\ f_0 \end{pmatrix} = \begin{pmatrix} -p \\ \phi \\ \phi_0 \end{pmatrix}. \tag{6.31}$$

In short notation we will write: $A f = \phi$.

6.4 Mapping Properties of the Operator Matrix

We introduce the product Sobolev spaces

$$\mathbf{H}_\pm^s := \begin{cases} [H^s(\Omega)]^3 \times [H^{s\pm1/2}(\Gamma)]^4 \times \mathbb{R}^1, & s \geq 0, \\ [\tilde{H}^s(\Omega)]^3 \times [H^{s\pm1/2}(\Gamma)]^4 \times \mathbb{R}^1, & s < 0 \end{cases}$$

and denote by $< . , . >$ the duality pairing between $\mathbf{H}_+^{-s}$ and $\mathbf{H}_-^s$.

Theorem 6.4.1. *For $-1/2 < s < 1$, the operator $\mathbb{A}$ is a continuous mapping from $\mathbf{H}_-^s$ to $\mathbf{H}_+^s$. For $s = 0$ holds Gårding's inequality:*

$$\langle \mathbb{A}\mathbf{f}, \mathbf{f} \rangle \geq C \left\| \mathbf{f}; \, \mathbf{H}_-^0 \right\|^2 - K(\mathbf{f}, \mathbf{f}), \qquad \forall \mathbf{f} \in \mathbf{H}_-^0.$$

Hereby C denotes a positive constant and $K(.,.)$ is a compact bilinear form in $\mathbf{H}_-^0$.

Proof. At first we consider the operator V^Γ and turn to an isothermal parametrization[1] of the mid-surface. Such a change of parametrization (in both directions) influences the boundary integral operators like a substitution of variables and the mapping properties are preserved.

In isothermal parametrization holds $\rho = |x - y|$ and the blocks $L^\square$ and $L_\square$ take the form

$$E(y) \left[I_{2\times 2} \ln \rho - \frac{\gamma}{\rho^2} \left(\begin{array}{cc} \vartheta_1^2 & \vartheta_1 \vartheta_2 \\ \vartheta_1 \vartheta_2 & \vartheta_2^2 \end{array} \right) \right], \tag{6.32}$$

where the additional pseudohomogeneous terms of degree ≥ 1 for the block $L_\square$ are neglected. The values of $E(y)$ are:

$$E(y) = \frac{(3 - \nu)\sqrt{a(y)}}{4\pi D(1 - \nu)}; \quad \gamma = \frac{1 + \nu}{3 - \nu}, \qquad \text{for } L^\square, \tag{6.33}$$

$$E(y) = \frac{\sqrt{a(y)}}{4\pi B}; \qquad \gamma = -1, \qquad \text{for } L_\square.$$

Detailed investigations for 2×2-systems of boundary integral operators with kernel functions of the form (6.32) acting on rectilinear polygons can be found in [26] and [27]. The second paper investigates especially the simple layer potential for the biharmonic operator, which has the form (6.32) with $\gamma = -1$. Theorems 5.1 and 5.2 from [27] assert that the corresponding operators are continuous mappings from $\left[H^{s-1/2}(\Gamma) \right]^2$ to $\left[H^{s+1/2}(\Gamma) \right]^2$ for $-1 < s < 1$ and that Gårding's inequality is valid for $s = 0$. These results can be carried over to the case $|\gamma| \leq 1$ [26] and also to regular curvilinear polygons [22]. This gives for V^Γ the mapping property

$$V^\Gamma : \left[H^{s-1/2}(\Gamma) \right]^4 \longmapsto \left[H^{s+1/2}(\Gamma) \right]^4, \qquad \forall s \in (-1, 1)$$

and for $s = 0$ we have Gårding's inequality:

$$\langle V^\Gamma f, f \rangle \geq C \|f\|_{-1/2}^2 - K'(f, f), \qquad \forall f \in \left[H^{-1/2}(\Gamma) \right]^4.$$

---

[1] The explicit knowledge of the isothermal parametrization is not necessary for the analytical investigations performed here.

Taking into accout, which minimal degree of pseudohomogeneity the individual components of the other operators have, we deduce from Theorems 3.7.1 and 3.7.3 the properties:

$$R^\Omega : \qquad [H^s(\Omega)]^3 \hookrightarrow [H^s(\Omega)]^3 \,, \qquad\qquad s \in \left(-\frac{1}{2}, \frac{3}{2}\right),$$

$$V^\Omega : \qquad [H^s(\Omega)]^3 \hookrightarrow \left[H^{s+1/2}(\Gamma)\right]^4 \,, \qquad s \in \left(-\frac{1}{2}, \frac{3}{2}\right),$$

$$R^\Gamma : \qquad \left[H^{s-1/2}(\Gamma)\right]^4 \hookrightarrow [H^s(\Omega)]^3 \,, \qquad s \in \left(-\frac{1}{2}, 1\right).$$

Now we take all pseudohomogeneous terms of degree 0 from $V^\Gamma(x, y)$ and denote the boundary integral operator with those kernel by V_0^Γ. Similary, R_{-1}^Γ denotes the operator whose kernel contains only the homogeneous terms of degree -1 from $R^\Gamma(x, y)$. The principal part of the operator matrix $\mathbb{A}$ can be written then:

$$\mathbb{A}_0 := \begin{pmatrix} I & -R_{-1}^\Gamma & 0 \\ 0 & V_0^\Gamma & e_0 \\ 0 & E_0 & 0 \end{pmatrix} : \mathbf{H}_-^s \mapsto \mathbf{H}_+^s, \quad \text{for} \quad -\frac{1}{2} < s < 1. \qquad (6.34)$$

The complete remainder $\mathbb{A} - \mathbb{A}_0$ is a compact pertubation of the operator (6.34). Gårding's inequality carries over from V_0^Γ to the complete operator $\mathbb{A}$ as can be deduced from the structure of $\mathbb{A}_0$. The proof is finished. ∎

Theorem 6.4.1 involves Fredholm's alternative for the operator $\mathbb{A}$. The example from section 3.2 shows that the invertibility can not be guaranteed. This must be assumed additionally.

For the practical realization of the BDIM, if necessary, this difficulty can be removed by modifications of the Levi function. As already mentioned, arbitrary smooth terms can be added to it.

Remark 6.4.1. The assertion of Theorem 6.4.1 holds as well for the adjoint operator $\mathbb{A}^*$.

This follows from the fact that the kernel functions belong to the same classes $\Psi\overline{\Omega}(\cdot)$ after changing the roles of x and y. Hence, they have the same mapping properties.

6.5 Corner Singularities

As already mentioned, the solution of the integral equation system may contain corner singularities even if the boundary data admit a smooth solution of the differential equations. As can be seen from (6.34), for smooth Dirichlet data $\phi(x)$, the leading singularities of the integral equation system are generated by the boundary operator V_0^Γ and then they are "transported" via

R^Γ_{-1} into the interior of the domain (i. e., onto the functions $w_j(x)$). More precisely this will be explained in Lemma 6.5.1.

At first we introduce a common notation for singular functions on Γ and on Ω:

$$\mathcal{V}[\lambda, l, \underline{x}] := \left\{ u(x - \underline{x}) \middle| : u(r, \varphi) = u_1(\varphi) r^{-i\lambda - 1} (\ln r)^l \right\}. \qquad (6.35)$$

The (complex) number λ is called the *exponent of the singular function*. Only the two (different) limit values of $u_1(\varphi)$ are of importance for singular functions on Γ. For singular functions on Ω, $u_1(\varphi)$ is a C^∞ function in the interval, which corresponds to the sector insight of Ω. Clearly, it can be continued as a periodic C^∞ function.

The singular functions belong to the following Sobolev spaces:

$$\mathcal{V}[\lambda, l, \underline{x}] \quad \subset \quad H^{\Im m\, \lambda - \varepsilon}(\Omega), \qquad (6.36)$$
$$\mathbf{tr}_x\, \mathcal{V}[\lambda, l, \underline{x}] \quad \subset \quad H^{\Im m\, \lambda - 1/2 - \varepsilon}(\Gamma). \qquad (6.37)$$

Both assertions hold for all $\varepsilon > 0$, all $l \in \mathbb{N}_0$ and for all boundary points $\underline{x} \in \Gamma$. (6.36) is a straightforward conclusion from (3.12). In case $\Im m\, \lambda > 1/2$, the assertion (6.37) can be derived by the well-known trace theorem. For $\Im m\, \lambda \leq 1/2$, the embedding must be verified directly by use of (3.16). But this case is not needed since it doesn't occur due to Remark 6.5.1.

We introduce the following spaces which contain the singular functions of interest. For an actual boundary value problem, $\mathcal{V}^s(\Gamma)$ and $\mathcal{V}^s(\Omega)$ denote the linear span of all (theoretically) occuring singular functions on Γ and Ω respectively, with exponents from the strip $0 < \Im m\, \lambda < s$. These spaces have finite dimensions.

The norm of a singular function from $\mathcal{V}^s(\Gamma)$ is the sum of the absolute values of both limits of $u_1(\varphi)$. For singular functions on Ω, the norm is defined by

$$\|u\|_{\mathcal{V}} := \left[\max_\varphi \left(|u_1(\varphi)|^2 + |u_1'(\varphi)|^2 \right) \right]^{1/2}. \qquad (6.38)$$

In case $0 < s < 1$, we denote by

$$\mathcal{V}^s \subset [\mathcal{V}^s(\Omega)]^3 \times [\mathcal{V}^s(\Gamma)]^4 \times \mathbb{R}^1$$

the product spaces which on each position contain the corresponding singular functions. The norm in $\mathcal{V}^s$ is the sum of the norms of all components.

Lemma 6.5.1. *Let Γ be a regular curvilinear polygonal and let G^Γ be the boundary integral operator (3.53) with a kernel $g(x, y) \in \Psi\overline{\Omega}(-1)$. Then G^Γ is a continuous mapping from $\mathcal{V}^s(\Gamma)$ to $\mathcal{V}^s(\Omega)$, for all $s \in (0, 1)$.*

Proof. The application of G^Γ to a singular function $u(x) \in \mathcal{V}[\lambda, l, \underline{x}]$ gives for $\Im m\, \lambda \in (0, 1)$ a function of the form

$$(G^\Gamma u)\,(x) = \sum_{\mu=0}^{l} b_\mu(x) u_\mu(x), \qquad u_\mu \in \mathcal{V}[\lambda,\mu,\underline{x}], \quad b_\mu(x) \in C^\infty$$

in a sufficiently small neighbourhood of the point $\underline{x}$. To show this, we put $\tau = i\lambda - 1$ and use in the remaining part of this proof the letter λ in other meaning, namely as the free variable of the Mellin transform. Without restriction of generality, we put $\underline{x} = 0$. Similar to the derivation of mapping properties, we start with the investigation of the operator (3.22) on the half-axis.

Let $\chi(x) \in C^\infty(\mathbb{R}_+)$ be a cut-off function with $\chi(x) \equiv 1$ for small x and $\chi(x) \equiv 0$ for large x. Its product with a singular function has the Mellin transform

$$\mathcal{M}\left[x^\tau(\ln x)^l \chi(x)\right](\lambda) = i(-i)^l \frac{d^l}{d\lambda^l}\left[\frac{\widehat{F}(\lambda - i\tau)}{\lambda - i\tau}\right], \qquad \Im \lambda < \Re \tau.$$

$\widehat{F}(\lambda)$ denotes the Mellin transform of the C_0^∞ function $x \cdot \chi'(x)$.

The Mellin transform of $(Gu)(x)$ in radial direction can be represented in the strip $-1 < \Im \lambda < \Re \tau$ in form of the product (3.24). The pole of order $l + 1$ at the point $\lambda = i\tau$ carries over to the product and it determines the type of the main singularity. Additional terms of order $\mu \leq l$ are contributed from the Taylor expansion of the function $\widehat{G}_\pm(\lambda, \phi)$ about the point $\lambda = i\tau$.

The passage to curvilinear polygons and the addition of pseudohomogeneous terms of degree ≥ 0 to the kernel $g(x,y)$ produces additional singular functions from $\mathcal{V}[\lambda + m, \mu, \underline{x}]$ with $m \in \mathbb{N}$ and $\mu \leq l$.   ∎

To determine the really occuring singular functions, we turn over again to an isothermal parametrization. (By the back-transformation to the actual parametrization, the type of the singularities is not changed.) Thereby, the polygon Γ goes over into a regular curvilinear polygonal Γ', which due to Lemma 6.1.1, has the same interiour angles like the mid-surface boundary. We denote the angles of Γ' by ω_j and the corner points of Γ by $\underline{x}^{(j)}$. The index j runs from 1 to J (= number of corners).

Then we have to calculate the zeros of the transcendental equation

$$\left[\gamma^2\lambda^2\sin^2(\omega) - \sinh^2(\omega\lambda)\right]\left\{\gamma^2\lambda^2\sin^2(2\pi - \omega) - \sinh^2[(2\pi - \omega)\lambda]\right\} = 0 \tag{6.39}$$

for all $\omega = \omega_j$ (see [26, Theorem 3.1] or [27, formula (4.5)]). Only solutions from the upper half-plane come into consideration, since for $\Im \lambda \leq 0$ the singular functions from $\mathcal{V}[\lambda, l, \underline{x}]$ don't belong to the "energetic space" $H^{-1/2}(\Gamma)$, i. e., these solutions are non-physical.

We denote the zeros of (6.39) by $\lambda_{jk}(\gamma)$ $(k = 1, 2, \ldots)$ and by $l_{jk}(\gamma)$ its multiplicity (which is 4 at most). Further,

$$I(0,1) := \{\Im \lambda_{jk}\} \cap (0,1)$$

denotes the set of the imaginary parts of all accuring exponents of singular functions which belong to the open interval $(0,1)$.

If we consider the operator $L^\square$, the value of γ is

$$\gamma = \gamma_\nu := \frac{1+\nu}{3-\nu} \in \left[\frac{1}{3}, \frac{3}{5}\right]$$

Singular functions from

$$\mathcal{V}\left[\lambda_{jk}(\gamma_\nu), l_{jk}(\gamma_\nu) - 1, \underline{x}^{(j)}\right]$$

are generated then on the positions f_1 and f_2 [26]. As explained in Lemma 6.5.1, the corresponding singular behaviour is carried over by R^Γ to the components w_j. For the operator $L_\square$, the number γ in (6.39) takes the value 1. Singular functions from

$$\mathcal{V}\left[\lambda_{jk}(1), l_{jk}(1) - 1, \underline{x}^{(j)}\right]$$

are generated then on the positions f_3 and f_4 [27]. Now R^Γ carries over singular functions with the exponents $\lambda_{jk}(1)$ to w_3. It is also carried over to w_1 and w_2, but here the exponent is raised by 1, as can be seen from (6.25).

Theorem 6.5.1. *Let* $\mathbf{A}$ *be an invertible mapping from* $\mathbf{H}_-^0$ *to* $\mathbf{H}_+^0$ *and let* s *be a number from* $I(0,1)$. *The following smoothness is assumed for the Dirichlet data and the surface loads:*

$$p^k(x) \in H^s(\Omega) \quad (k = 1, 2, 3); \qquad \phi_k(x) \in H^{s+1/2}(\Gamma) \quad (k = 1, 2, 3, 4).$$

Then the operator $\mathbf{A}$ *is an invertible mapping from*

$$\mathcal{V}\mathbf{H}_-^s := \mathcal{V}^s \oplus \mathbf{H}_-^s \tag{6.40}$$

to $\mathbf{H}_+^s$ *and the solution of the equation* (6.31) *is composed from a regular and a singular part:*

$$\mathbf{f} = \mathbf{f}_{\mathrm{reg}} + \mathbf{f}_{\mathrm{sing}}; \qquad \mathbf{f}_{\mathrm{reg}} \in \mathbf{H}_-^s, \quad \mathbf{f}_{\mathrm{sing}} \in \mathcal{V}^s.$$

This theorem can be deduced from [27, Theorem 5.5] and Lemma 6.5.1. With (6.36) and (6.37) we further obtain:

Corollary 6.5.1. *Denote* $\lambda_{\min} := \min(\Im \lambda_{j,k})$, *where we have to minimize over all solutions of* (6.39) *which lay in the upper half-plane. If the assumptions of Theorem 6.5.1 are fulfilled, then the solution of* (6.31) *belongs to the space* $\mathbf{H}_-^s$ *for all* $s < \lambda_{\min}$.

Remark 6.5.1. For $|\gamma| \leq 1$ and $\omega \in (0, 2\pi)$, all solutions λ of (6.39) from the upper half-plane satisfy the inequality $\Im \lambda > 1/2$. Thus the solution of (6.31) always belongs to the space $\mathbf{H}_-^{1/2}$ for sufficiently smooth right-hand sides.

6.6 Galerkin's Method

The meshwidth of the discretization is denoted by h. It is assumed that h tends to zero (in discrete steps). For each value of h we have a (disjunct) subdivision

$$\overline{\Omega} = \bigcup_{k=1}^{N(h)} \overline{\Omega}_k^{(h)}$$

of the domain and a (disjunct) subdivision

$$\overline{\Gamma}_j = \bigcup_{k=1}^{N_j(h)} \overline{\Gamma}_{jk}^{(h)}, \qquad j = 1, \dots, J$$

of each boundary segment. Thereby holds:

$$diam\ \Omega_k^{(h)} \le h; \quad \left|\Gamma_{jk}^{(h)}\right| \le h; \quad N(h) = \mathcal{O}(h^{-2}); \quad N_j(h) = \mathcal{O}(h^{-1}).$$

For the domain subdivision, we require that no very long and very thin elements occur. Precisely formulated, a constant $\bar{c} > 0$ must exist, with

$$\left|\Omega_k^{(h)}\right| := meas\ \Omega_k^{(h)} \ge \bar{c} \left[diam\ \Omega_k^{(h)}\right]^2, \qquad \forall k, \forall h. \qquad (6.41)$$

Further requirements concerning the form and the topological relationships of the elements are not needed.

For numerical solution of the integral equation system (6.31), Galerkin's method is used with following trial functions:

– piecewise constant functions in Ω,
– "hat functions" on Γ, which are piecewise linear and continuous even at the corner points.

The finite-dimensional spaces spanned by these trial functions are denoted by

$$\mathbf{S}_h := \left[S_h^0(\Omega)\right]^3 \times \left[S_h^1(\Gamma)\right]^4 \times \mathbb{R}^1.$$

These are subspaces of $\mathbf{H}_+^{1-\varepsilon}$ for all $\varepsilon > 0$. Its dimension amounts to

$$dim\ \mathbf{S}_h = 3N(h) + 4L(h) + 1 = \mathcal{O}(h^{-2}) \qquad (6.42)$$

with

$$L(h) := \sum_{j=1}^{J} N_j(h) = \mathcal{O}(h^{-1}).$$

Of course, it is possible to use smoother splines as well. In our numerical experiments we used only the trial functions mentioned above. For simplicity, we restrict our investigations to this case.

Galerkin's method, as well-known, consists in the determination of an approximate solution $f_h \in S_h$, which satisfies

$$\langle A\, f_h, \psi_h \rangle = \langle \phi, \psi_h \rangle, \qquad \forall \psi_h \in S_h. \tag{6.43}$$

This leads to fully populated matrices of dimension (6.42). The calculation of each matrix entry requires the numerical evaluation of a multiple integral and this is very time consuming if it is done as exactly as possible. In order to save computation time, it is appropriate to calculate these multiple integrals only with a well-adapted reduced accuracy instead of the highest reachable one. The exact bilinear form $< A \cdot, \cdot >$ from (6.43) is replaced by an *approximate bilinear form* $< \tilde{A} \cdot, \cdot >$. We claim that the order of convergence is not disturbed, if the following condition is satisfied:

$$\left| \left\langle (A - \tilde{A}) \psi_h, \psi_h' \right\rangle \right| \leq C_K\, h^2\, \|\psi_h\|_0\, \|\psi_h'\|_0, \qquad \forall \psi_h, \psi_h' \in S_h. \tag{6.44}$$

Here and in what follows, we denote by C_K a generic constant and by $\|.\|_0$ the norm in H^0_-.

The real generated approximate solution comes from the disturbed linear system

$$\left\langle \tilde{A}\, \tilde{f}_h, \psi_h \right\rangle = \langle \phi, \psi_h \rangle, \qquad \forall \psi_h \in S_h. \tag{6.45}$$

We assume, that the right-hand side is discretized exactly, though here an approximation error is allowed as well, namely an error of order $h^2 \|\psi_h\|_0$. For simplicity, this is not taken into account, but it can be included into the error estimates without difficulties.

6.7 Error Estimates

Gårding's inequality from Theorem 6.4.1 is equivalent to the fact, that the operator A can be written as the sum of a positive definit and a compact operator. Furthermore, from well-known properties of trial functions we can deduce that the orthoprojections from H^0_- to S_h converge strongly to the identity for $h \to 0$. Under these assumptions, as was shown in [61], a mesh-width $h_0 > 0$ and a constant $C > 0$ exists such that for all $h < h_0$ the "condition of uniform S_h-ellipticity"

$$|\langle A\psi_h, \psi_h \rangle| \geq C \|\psi_h\|_0^2, \qquad \forall \psi_h \in S_h$$

is fulfilled. Then, due to the theorem of Lax and Milgram, the linear system (6.43) is uniquely solvable for all $h < h_0$. With Céa's lemma [18, Theorem 2.4.1], we obtain the error estimate

$$\|f - f_h\|_0 \leq C \inf_{\psi_h \in S_h} \|f - \psi_h\|_0. \tag{6.46}$$

The letter C stands here and in the sequel for an arbitrary positive constant which does not depend on h and the considered functions. From step to step this can be an other constant.

From (6.44) follows, that the disturbed bilinear form also fulfills the condition of uniform S_h-ellipticity for h sufficiently small. Thus, the first Strang Lemma [18, Theorem 4.1.1] can be applied, which asserts in our case:

$$\left\| \mathbf{f} - \tilde{\mathbf{f}}_h \right\|_0 \le C \inf_{\psi_h \in S_h} \left[\left\| \mathbf{f} - \psi_h \right\|_0 + \sup_{\psi'_h \in S_h} \frac{\left| \left\langle (\mathbf{A} - \tilde{\mathbf{A}}) \psi_h, \psi'_h \right\rangle \right|}{\left\| \psi'_h \right\|_0} \right]. \quad (6.47)$$

The second term in the square bracket can be estimated from above by $C_K h^2 \|\psi_h\|_0$. With the triangular inequality

$$\|\psi_h\|_0 \le \|\mathbf{f} - \psi_h\|_0 + \|\mathbf{f}\|_0,$$

we obtain (with a new constant) the error estimate

$$\left\| \mathbf{f} - \tilde{\mathbf{f}}_h \right\|_0 \le C \left(\inf_{\psi_h \in S_h} \|\mathbf{f} - \psi_h\|_0 + h^2 \|\mathbf{f}\|_0 \right). \quad (6.48)$$

As can be seen, the effective error depends on the accuracy of the best possible approximation of $\mathbf{f}$ by the trial functions. This is the crucial point which restricts the reachable order of convergence.

It is possible to apply the Aubin-Nitsche lemma to achieve a higher order of convergence in a weaker norm. The (probably) first application of this method to integral equations was given in [68]. One needs the assumption that the assertions about invertibilty and mapping properties from Theorem 6.5.1 hold as well for the adjoint operator $\mathbf{A}^*$. If $\mathbf{A}$ is invertible, this is ensured due to Remark 6.4.1.

In the case $s \in I(0, 1)$, a solution of the equation $\mathbf{A}^* \mathbf{v} = \mathbf{w}$ exists for all $\mathbf{w} \in \mathbf{H}^s_+$ and the inequality

$$\left\| \mathbf{v}; \mathcal{V} \mathbf{H}^s_- \right\| \le C_A \left\| \mathbf{w}; \mathbf{H}^s_+ \right\|$$

holds, where C_A is the norm of the inverse operator and $\mathcal{V} \mathbf{H}^s_-$ is the space defined by (6.40). The error of our numerical solution can be estimated:

$$\left\| \mathbf{f} - \tilde{\mathbf{f}}_h; \mathbf{H}^{-s} \right\| = \sup_{\mathbf{w} \in \mathbf{H}^s_+} \frac{\left| \left\langle \mathbf{f} - \tilde{\mathbf{f}}_h, \mathbf{w} \right\rangle \right|}{\|\mathbf{w}; \mathbf{H}^s_+\|} =$$

$$= \sup_{\mathbf{v} \in \mathcal{V} \mathbf{H}^s_-} \frac{\left| \left\langle \mathbf{f} - \tilde{\mathbf{f}}_h, \mathbf{A}^* \mathbf{v} \right\rangle \right|}{\|\mathbf{A}^* \mathbf{v}; \mathbf{H}^s_+\|} \le C_A \sup_{\mathbf{v}} \frac{\left| \left\langle \mathbf{A}(\mathbf{f} - \tilde{\mathbf{f}}_h), \mathbf{v} \right\rangle \right|}{\|\mathbf{v}; \mathcal{V} \mathbf{H}^s_-\|}.$$

Subtracting (6.45) from (6.43) gives

$$\left\langle \mathbf{A}\,\mathbf{f} - \tilde{\mathbf{A}}\,\tilde{\mathbf{f}}_h, \boldsymbol{\psi}_h \right\rangle = 0, \qquad \forall \boldsymbol{\psi}_h \in \mathbf{S}_h.$$

For all test functions $\boldsymbol{\psi}_h \in \mathbf{S}_h$ holds

$$\left\langle \mathbf{A}(\mathbf{f} - \tilde{\mathbf{f}}_h), \mathbf{v} \right\rangle =$$

$$= \left\langle \mathbf{A}(\mathbf{f} - \tilde{\mathbf{f}}_h), \mathbf{v} - \boldsymbol{\psi}_h \right\rangle + \left\langle (\mathbf{A} - \tilde{\mathbf{A}})\tilde{\mathbf{f}}_h, \mathbf{v} - \boldsymbol{\psi}_h \right\rangle + \left\langle (\tilde{\mathbf{A}} - \mathbf{A})\tilde{\mathbf{f}}_h, \mathbf{v} \right\rangle.$$

We still introduce the notation

$$K(s,h) := \sup_{\mathbf{v} \in \mathcal{V}\mathbf{H}_-^s} \inf_{\boldsymbol{\psi}_h \in \mathbf{S}_h} \frac{\|\mathbf{v} - \boldsymbol{\psi}_h\|_0}{\|\mathbf{v};\, \mathcal{V}\mathbf{H}_-^s\|}. \tag{6.49}$$

From (6.44), (6.48) and the inequality $\|\mathbf{v}\|_0 \leq \|\mathbf{v};\, \mathcal{V}\mathbf{H}_-^s\|$, we obtain the error estimate

$$\left\|\mathbf{f} - \tilde{\mathbf{f}}_h;\, \mathbf{H}_-^{-s}\right\| \leq$$

$$\leq C_A \left[\left(\|\mathbf{A};\, \mathbf{H}_-^0 \mapsto \mathbf{H}_+^0\| \left\|\mathbf{f} - \tilde{\mathbf{f}}_h\right\|_0 + C_K h^2 \left\|\tilde{\mathbf{f}}_h\right\|_0 \right) K(s,h) + C_K\, h^2 \left\|\tilde{\mathbf{f}}_h\right\|_0 \right]$$

$$\leq C \left\{ [K(s,h)]^2 + h^2 \right\} \|\boldsymbol{\phi}\|_s \tag{6.50}$$

with

$$\|\boldsymbol{\phi}\|_s := \left\|\boldsymbol{\phi};\, \mathbf{H}_+^s\right\| \geq \left(\|\mathbf{A}^{-1};\, \mathbf{H}_+^s \mapsto \mathcal{V}\mathbf{H}_-^s\| \right)^{-1} \left\|\mathbf{f};\, \mathcal{V}\mathbf{H}_-^s\right\|.$$

6.8 Order of Convergence for Quasi-Uniform Meshes

The subdivision is called *quasi-uniform*, if a positive constant $\underline{c}$, independent of h, exists such that

$$diam\ \Omega_k^{(h)} \geq \underline{c}\, h; \qquad \left|\Gamma_{jk}^{(h)}\right| \geq \underline{c}\, h.$$

If additionally, condition (6.41) is fulfilled, then the best approximation of functions $w(x) \in H^s(\Omega)$ by piecewise constant trial functions satisfies the estimate [36, Theorem 2.27]:

$$\inf_{w_h \in S_h^0(\Omega)} \left\|w - w_h;\, H^0(\Omega)\right\| \leq C h^s \left\|w;\, H^s(\Omega)\right\|, \qquad 0 \leq s \leq 1. \tag{6.51}$$

A general theory for the approximation properties of splines in Sobolev spaces is worked out in [6]. Its consequences for the special case of H^s-spaces on closed, smooth curves (equivalently, for Sobolev spaces of periodic functions), can be found in [5], [34, sect. 6.1] and [67]. These results can be carried over by use of (3.16) to the spaces $H^s(\Gamma)$ considered here, provided that the corners of the polygon belong to the set of discretization points. We obtain

$$\inf_{f_h \in S_h^1(\Gamma)} \left\| f - f_h; \, H^{-1/2}(\Gamma) \right\| \le C h^s \left\| f; \, H^{s-1/2}(\Gamma) \right\|, \qquad 0 \le s \le 5/2.$$

(6.52)

If $\mathbf{v}$ is an arbitrary vector from the space $\mathbf{H}_-^s$, it can be approximated with the accuracy:

$$\inf_{\psi_h \in S_h} \left\| \mathbf{v} - \psi_h; \, \mathbf{H}_-^0 \right\| \le C h^s \left\| \mathbf{v}; \, \mathbf{H}_-^s \right\|, \qquad 0 \le s \le 1.$$

Due to Conclusion 6.5.1, the solution of our system of integral equations belongs to the space $\mathbf{H}_-^s$ for $s < \lambda_{\min}$. In this case (and only in this case), the spaces $\mathcal{V}\mathbf{H}_-^s$ and $\mathbf{H}_-^s$ coincide. In general the value of $\lambda_{\min}$ is smaller than 1. From (6.46) we obtain the following order of convergence for the Galerkin solution:

$$\left\| \mathbf{f} - \tilde{\mathbf{f}}_h \right\|_0 \le C \|\phi\|_s \, h^s, \qquad \text{for} \quad s \in [0,1); \quad s < \lambda_{\min}.$$

(6.53)

From (6.50) follows:

$$\left\| \mathbf{f} - \tilde{\mathbf{f}}_h; \, H^{-s}_- \right\| \le C \|\phi\|_s \, h^{2s}, \qquad \text{for} \quad s \in [0,1); \quad s < \lambda_{\min}.$$

(6.54)

This enhanced order of convergence in a weaker norm seems to be of advantage when the displacement field is calculated a-posteriori by inserting the approximate solution $\tilde{f}_h$ into (6.21). As can be seem from (6.22) and (6.23), the components of $L^\Omega(x,y)$ and $L^\Gamma(x,y)$ belong to the classes $\Psi\overline{\Omega}(0)$ or they are smoother. Applying Theorems 3.7.1 and 3.7.3, we obtain for $0 \le s < 1$ and $s < \lambda_{\min}$:

$$\left\| L^\Omega(w - \tilde{w}_h); \, [H^{2-s}(\Omega)]^3 \right\| \le C \|\phi\|_s \, h^{2s},$$

$$\left\| L^\Gamma(f - \tilde{f}_h); \, [H^{1-s}(\Omega)]^3 \right\| \le C \|\phi\|_s \, h^{2s}.$$

Collecting all results now, we obtain an error estimate for the approximate solution $\tilde{v}_h := L^\Omega \tilde{w}_h + L^\Gamma \tilde{f}_h$:

$$\left\| v - \tilde{v}_h; \, [H^{1-s}(\Omega)]^3 \right\| \le C h^{2s} \left(\left\| p; \, [H^s(\Omega)]^3 \right\| + \left\| \phi; \, \left[H^{s+1/2}(\Gamma) \right]^4 \right\| \right)$$

(6.55)

which holds for $0 \le s < 1$; $s < \lambda_{\min}$. Clearly, the left side of (6.55) can be replaced by a weaker norm, especially the L^2 norm.

6.9 Approximation of the Singular Functions on Graded Meshes

The convergence order obtained in (6.55) is not satisfactory. In cases of practical interest, the exponent $2\lambda_{\min}$ may be only less larger than 1 (see section

7.5). The singular functions disturb a faster convergence. (This is a special kind of a "pollution effect".) Two possibilities are known to achieve better convergence. The first are the so called "augmented methods", where the singular functions are added to the space of the trial functions. Augmented Galerkin methods for boundary integral equations are investigated, e. g., in [24] and [26].

The second possibility consists in graded meshes with a well-adapted refinement in the vicinity of the corners. This method was realized in our numerical experiments, since the implementation is easier than for augmented methods. The first step consists hereby in the determination of a refinement parameter $q_j \geq 1$ for each corner point. Regarding this, we refer to Lemma 6.9.1 and Lemma 6.9.2. If q_j is fixed, the mesh is refined in the vicinity of the corresponding corner to the q_jth power of the distance. What is to be understood under this terminology, will be explained at first for the interval $(0,1)$. Let M be the number of subintervalls. For a refinement to the qth power in the vicinity of zero, the endpoints of the subintervalls are choosen as: $t_0 = 0$; $t_k = (k/M)^q$ $(k = 1, \ldots, M)$ (cmp. [104, Lemma 1]).

A corresponding refinement of Γ can be constructed by dividing each boundary segment into halfs. Each of them is mapped to the interval $(0,1)$ by a linear transformation with respect to the line parameter. The back-transformed endpoints of the $N_j(h)/2$ subintervalls (with corresponding refinement) give the discretization points on the boundary (cmp. [105, formula (3.1)]).

A refinement of the domain discretization to the qth power is characterized by the following properties [4, sect. 3.1] (see also [33, sect. 2]):

1. Without restriction of generality, let $\Omega_1^{(h)}$ be the only element touching the actual corner point. For its diameter we require: $h_1 := diam \, \Omega_1^{(h)} = \mathcal{O}(h^q)$.

2. The distance of the domain $\Omega \setminus \Omega_1^{(h)}$ to the corner point, denoted by h_0, is of the same order as h_1.

3. For all elements not touching the corner point, the expression

$$\frac{h \left| x_{(k)} \right|^{1-1/q}}{h_k}$$

is uniformly bounded from above and below. Hereby, h_k denotes the diameter of the element and $x_{(k)}$ denotes its centroid.

Now we investigate the approximation of singular functions on graded meshes.

Lemma 6.9.1. *We consider a singular function from* $u \in \mathcal{V}[\lambda, l, \underline{x}]$ *with* $1/2 < \Im \lambda \leq 1$. *If the discretization is refined in the vicinity of the corner point* $\underline{x} \in \Gamma$ *to the qth power, the best approximation with hat functions satisfies:*

$$\inf_{u_h \in S_h^1(\Gamma)} \left\| u - u_h; \, H^{-1/2}(\Gamma) \right\| \le C \begin{cases} h^Q \, |\ln h|^l, & \text{for } Q < 2, \\ h^2 \, |\ln h|^{l+1/2}, & \text{for } Q = 2, \\ h^2, & \text{for } Q > 2, \end{cases}$$

with $Q = q \, \Im \lambda$.

Proof. We consider the case $l = \Re \lambda = 0$ and investigate the function x^{p-1} for $p = \Im \lambda$ on the interval $[0, 1]$. An asymptotic optimal approximation can be achieved with a piecewise linear function $u_h(x)$ which takes the value 0 for $x = 0$ and which takes the values t_k^p at the nodal points $t_k = (k/M)^q$. The length of the largest interval is $h = 1 - t_{M-1} = \mathcal{O}(M^{-1})$. The lengths of the other intervals are

$$h_k := t_{k+1} - t_k = t_k \left[\left(1 + \frac{1}{k} \right)^q - 1 \right] = t_k \left[\frac{q}{k} + \mathcal{O}(k^{-2}) \right]. \tag{6.56}$$

From [14, formula 3.14] the following estimate can be taken, which holds for arbitrary subdivisions with $0 = t_0 < t_1 < \ldots < t_M = 1$:

$$\left\| x^{p-1} - u_h; \, \tilde{H}^{-1/2}(0,1) \right\|^2 \le C \left[t_1^{2p} + \sum_{k=1}^{M-1} \left(\frac{h_k}{t_k} \right)^5 t_k^{2p} \right].$$

With (6.56), this can be estimated for the special choice $t_k = (k/M)^q$ from above:

$$\cdots \le C h^{2pq} \left(1 + \sum_{k=1}^{M-1} k^{2pq-5} \right) \le C \begin{cases} h^{2pq}, & pq < 2; \\ h^{2pq} |\ln h|, & pq = 2; \\ h^4, & pq > 2. \end{cases}$$

Here we use [14, Lemma 3.6], which asserts that the series is bounded by $C M^{2pq-4}$ in the case $pq > 2$.

From the obtained estimate on the interval $(0, 1)$, a corresponding estimate on each segment Γ_j can be derived by the linear transformation mentioned above. The assertion for the complete boundary can be verified then with [14, Lemma 3.1], which supplies the inequality

$$\left\| u; \, H^{-1/2}(\Gamma) \right\| \le \left(\sum_{j=1}^{J} \left\| u; \, \tilde{H}^{-1/2}(\Gamma_j) \right\| \right)^{1/2}$$

for each function $u(x) \in L^2(\Gamma)$. Thus the lemma is proved for $l = 0$. The estimates for $l > 0$ are more sophisticated, but follow the same sceme. ∎

The approximation of singular functions on polygons was already investigated in [104] for continuous (not continuously differentiable) splines and in [105] for the class of smoothest splines. In the special case of hat functions, both papers provide only the result, the order h^2 of approximation can be achieved when the refinement parameter is choosen as $q = 5/(2 \Im \lambda - 1)$. The assertion of Lemma 6.9.1 for $Q > 2$ is contained in [14]. Only such refinements are considered there, which lead to the approximation order h^2.

For the practical realization of the method, however, it is essential to choose, if possible, the minimal refinement parameter which ensures still the desired order of approximation. The larger the refinement parameter is, the larger is the condition number of the discretization matrix [14] and the larger are the numerical problems originating from necessary summations of real numbers with significantly different magnitude, as they accur during the generation and the solution of the discretized systems.

In our case it suffices to put the constant Q from Lemma 6.9.1 equal to 1. The reachable order of convergence is restricted already due to the mapping properties of the complete operator and the domain discretization. It makes no sense, to perform the approximation on the boundary more accurately than necessary, i. e., to produce an error here of lower order than elsewhere.

Lemma 6.9.2. *Consider a singular function $u \in V[\lambda, l, \underline{x}]$ with $0 < \Im \lambda \leq 1$ in the domain Ω. If we put $q = Q/\Im \lambda$ and refine the domain in the vicinity of the corner point $\underline{x} \in \Gamma$ to the q th power, the best approximation by piecewise constant functions fulfills:*

$$\inf_{w_h \in S_h^0(\Omega)} \|u - w_h\|_{L^2(\Omega)} \leq C \|u_1\|_\nu \begin{cases} h^Q |\ln h|^l, & \text{for } \Im \lambda \leq Q < 1, \\ h |\ln h|^{l+1/2}, & \text{for } Q = 1, \\ h, & \text{for } Q > 1. \end{cases}$$

The norm $\|u_1\|_\nu$ is defined by (6.38).

Proof. Put $\underline{x} = 0$. The proof is only performed for the case $l = \Re \lambda = 0$. The other cases can be handled analogically.

We consider the singular function $u(r, \phi) = r^p u_1(\phi)$ with $p = \Im \lambda - 1$ and investigate its approximation by a piecewise constant function w_h, defined as follows. On $\Omega_1^{(h)}$, w_h equals zero; else it is equal to the value of the function in the centroid. The expression for $\|u - w_h\|_{L^2(\Omega)}^2$ is split into the sum of integrals on the individual elements.

The integral over $\Omega_1^{(h)}$ is less or equal to

$$\int_{|x| \leq h_1} r^{2p} [u_1(\varphi)]^2 dx \leq \frac{h_1^{2(p+1)}}{2(p+1)} \|u_1(\varphi)\|_{L^2}^2 \leq C \|u_1(\varphi)\|_{L^2}^2 h^{2Q}.$$

In the other integrals, the integrand is replaced by the first term of the corresponding Taylor series, taken to the square. The sum on the domain $\Omega \setminus \Omega_1^{(h)}$ is bounded from above by

$$C \max_\varphi \left[|u_1'(\varphi)|^2 + |u_1(\varphi)|^2\right] \sum_{k=2}^{N(h)} \left|\Omega_k^{(h)}\right| \left|x_{(k)}\right|^{2(p-1)} h_k^2 \leq$$

$$\leq C h^2 \|u_1\|_\nu^2 \sum_{k=2}^{N(h)} \left|\Omega_k^{(h)}\right| \left|x_{(k)}\right|^{2(p-1/q)}.$$

In the last step we have replaced h_k by $h|x_{(k)}|^{1-1/q}$, due to the third property of the domain subdivision. The obtained expression can be estimated further by

$$\sum_{k=2}^{N(h)} \int_{\Omega_k^{(h)}} |x|^{2(z-1)} dx = \sum_{k=2}^{N(h)} \left[|\Omega_k^{(h)}| \, |x_{(k)}|^{2(z-1)} + \mathcal{O}\left(h_k^4\right) \right]$$

$$\leq \int_{|x| \geq h_0} |x|^{2(z-1)} dx \leq C \begin{cases} h_0^{2z}, & z < 0, \\ \ln h_0, & z = 0, \\ 1, & z > 0, \end{cases}$$

which is valid for all $z \in \mathbb{R}^1$. Taking $z = 1 + p - 1/q = (Q-1)/q$, we obtain the assertion of Lemma 6.9.2. ∎

6.10 Order of Convergence for Graded Meshes

Assume, the right side of equation (6.31) belongs to the space $\mathbf{H}_+^1$. Then Theorem 6.5.1 implies $\mathbf{f} \in \mathcal{V}\mathbf{H}_-^s$ for all $s \in I(0,1)$.

The constant Q from Lemma 6.9.1 and Lemma 6.9.2 is set to 1. This means, the refinement parameter q_j assigned to the corner point $\underline{x}^{(j)}$ is determined by

$$q_j \geq \min_k \left\{ \frac{1}{\mathfrak{Im} \, \lambda_{jk}} \, \Big| : \mathfrak{Im} \, \lambda_{jk} > 1/2 \right\}. \tag{6.57}$$

It can be proved that the first solution of equation (6.39) (e. g., the solution with the smallest positive imaginary part) has the multiplicity 1. Thus, the corresponding order of convergence doesn't contain a logarithmic term. Due to Lemma 6.9.1, all singular functions on Γ are then approximated with the order h. The regular part of the solution can be estimated by (6.52). Summarily we obtain the following estimate for the components $f \in \{f_1, f_2, f_3, f_4\}$:

$$\inf_{f_h \in S_h^1(\Gamma)} \left\| f - f_h; \, H^{-1/2}(\Gamma) \right\| \leq C \, h^s \left\| f; \, H^{s-1/2}(\Gamma) \right\|, \quad s \in I(0,1). \tag{6.58}$$

A similar result is given in [105, Theorem 3.1], but with a larger refinement parameter.

For the components $w \in \{w_1, w_2, w_3\}$ we obtain from Lemma 6.9.2 and from (6.51) the estimate:

$$\inf_{w_h \in S_h^0(\Omega)} \left\| w - w_h \right\|_{L^2(\Omega)} \leq C \, h^s \left\| w; \, H^s(\Omega) \right\|; \quad s \in I(0,1). \tag{6.59}$$

This leads to the inequality $K(s,h) \leq C \, h^s$ for $s \in I(0,1)$, where $K(s,h)$ is the expression (6.49). Inserting this into the error estimates (6.48) and (6.50), we get the following final result.

Theorem 6.10.1. *Suppose that the assumptions from Theorem 6.5.1 are fulfilled, that the refinement parameters for each corner point are choosen according to formula (6.57) and that condition (6.44) is satisfied for the approximate calculation of the matrix entries. Then the linear system (6.45) is uniquely solvable for all sufficiently small h. There exists a constant C (only depending on s and Ω), such that the Galerkin solution of the complete integral equation system fulfills the error estimate*

$$\left\| \mathbf{f} - \tilde{\mathbf{f}}_h \right\|_0 \leq C h^s \|\phi\|_s,$$

$$\left\| \mathbf{f} - \tilde{\mathbf{f}}_h; \mathbf{H}_-^{-s} \right\| \leq C h^{2s} \|\phi\|_s$$

for all $s \in I(0,1)$. The estimate (6.55) for the error of the displacement field holds now for all $s \in I(0,1)$.

The set $I(0,1)$ consists of the intervall $(0,1)$ with the exception of finitely many points. For a right-hand side of $\mathbf{H}_+^1$, we almost obtain h^2 as order of convergence for the displacement field. This order was confirmed by our numerical experiments (see section 7.8).

6.11 Estimation of the Admissible Local Quadrature Error

Inequality (6.44) has global character since it must be fulfilled for all functions of $\mathbf{S}_h$. The question arises, which accuracy must be required for the particular entries of the discretization matrices to ensure that (6.44) is fulfilled.

Similar considerations for the FEM can be found in [18, sect. 4.1]. But here we have other operators and we need estimates in $H^{-1/2}(\Gamma)$. That is why we will outline the corresponding assertions for integral operators in the following. Thereby the letter B stands for the particular entries of the matrix $A - \tilde{A}$. Clearly, these relations hold as well for arbitrary linear continuous operators B, for which the corresponding duality pairings make sense.

Let

$$g(y) = \sum_{k=1}^{N(h)} g_k \chi_k^\Omega(y) \in S_h^0(\Omega)$$

be a piecewise constant function. By χ_k^Ω we denote the characteristic function of $\Omega_k^{(h)}$. From the well-known inequality

$$\left(\sum_{k=1}^N a_k \right)^2 \leq N \sum_{k=1}^N a_k^2 \tag{6.60}$$

we obtain

$$\|g(y); L^2(\Omega)\| = \left(\sum_{k=1}^{N(h)} g_k^2 \left|\Omega_k^{(h)}\right|\right)^{1/2} \geq \frac{1}{\sqrt{N(h)}} \sum_k |g_k| \left|\Omega_k^{(h)}\right|^{1/2}.$$

Further, let

$$g'(y) = \sum_{k=1}^{N(h)} g_k' \chi_k^\Omega(y) \in S_h^0(\Omega)$$

be a second piecewise constant function. Then

$$|< Bg, g' >| = \left|\sum_k \sum_j g_k \, g_j' \left\langle B\chi_k^\Omega, \chi_j^\Omega \right\rangle\right| \leq$$

$$\leq N(h) \max_{k,j} \frac{\left|\langle B\chi_k^\Omega, \chi_j^\Omega \rangle\right|}{\left(\left|\Omega_k^{(h)}\right|\left|\Omega_j^{(h)}\right|\right)^{1/2}} \, \|g; L^2(\Omega)\| \, \|g'; L^2(\Omega)\|.$$

From $N(h) = \mathcal{O}(h^{-2})$ we conclude that the local condition

$$\left|\langle B\chi_k^\Omega, \chi_j^\Omega \rangle\right| \leq Ch^4 \left(\left|\Omega_k^{(h)}\right|\left|\Omega_j^{(h)}\right|\right)^{1/2}, \qquad \forall k,j \qquad (6.61)$$

is sufficient for the global estimate

$$|\langle Bg, g' \rangle| \leq Ch^2 \|g; L^2(\Omega)\| \, \|g'; L^2(\Omega)\|, \qquad \forall g, g' \in S_h^0(\Omega)$$

still to be fulfilled. For a quasi-uniform mesh, the right-hand side of (6.61) can be replaced by Ch^6.

Let

$$f(y) = \sum_{k=1}^{L(h)} f_k \chi_k^\Gamma(y) \in S_h^1(\Gamma)$$

be a piecewise linear function on the boundary Γ, where χ_k^Γ denote the hat functions. With the notation $h_k := meas \ supp \ \chi_k^\Gamma$ for the local meshwidth, we obtain by use of (6.60) the inequality

$$\|f(y); L^2(\Gamma)\| = \left[\sum_{k=1}^{L(h)} f_k^2 \|\chi_k^\Gamma; L^2(\Gamma)\|^2 + \sum_{j \neq k} f_k \, f_j \langle \chi_k^\Gamma, \chi_j^\Gamma \rangle\right]^{1/2}$$

$$\geq C \left(\sum_{k=1}^{L(h)} f_k^2 h_k\right)^{1/2} \geq \frac{C}{\sqrt{L(h)}} \sum_k |f_k| \sqrt{h_k}.$$

To estimate the norm in $H^{-1/2}(\Gamma)$, we need the "inverse property"

$$\left\|f; H^{-1/2}(\Gamma)\right\| \geq C \sqrt{h_{min}} \, \|f; L^2(\Gamma)\|$$

where $h_{\min}$ denotes the minimal meshwith on Γ. Such a result has been stated in [5] and [34, Theorem 6.1.3] for certain scales of Sobolev spaces and for quasi-uniform meshes. Taking $h_{\min}$ instead of h, this estimate is true as well for graded meshes. (Exacter estimates with a certain mean value of h are possible, but more complicated.)

Let

$$f'(y) = \sum_{k=1}^{L(h)} f'_k \chi_k^\Gamma(y) \in S_h^1(\Gamma)$$

be a second piecewise linear function. Then

$$|< Bf, f' >| = \left| \sum_k \sum_j f_k \, f'_j \langle B\chi_k^\Gamma, \chi_j^\Gamma \rangle \right| \le$$

$$\le \; CL(h) \, h_{\min}^{-1} \max_{k,j} \frac{|\langle B\chi_k^\Gamma, \chi_j^\Gamma \rangle|}{\sqrt{h_k h_j}} \; \|f; \, H^{-1/2}(\Gamma)\| \, \|f'; \, H^{-1/2}(\Gamma)\|.$$

Since $L(h) = \mathcal{O}(h^{-1})$, the local estimate

$$|\langle B\chi_k^\Gamma, \chi_j^\Gamma \rangle| \le Ch^3 h_{\min} \sqrt{h_k h_j} \tag{6.62}$$

guarantees the global condition

$$|\langle Bf, f' \rangle| \le Ch^2 \, \|f; \, H^{-1/2}(\Gamma)\| \, \|g'; \, H^{-1/2}(\Gamma)\|, \qquad \forall f, f' \in S_h^1(\Gamma).$$

Analogously we conclude from the local estimations

$$\left. \begin{array}{c} |\langle B\chi_k^\Gamma, \chi_j^\Omega \rangle| \\ |\langle B\chi_j^\Omega, \chi_k^\Gamma \rangle| \end{array} \right\} \le Ch^{7/2} \sqrt{h_{\min}} \left(h_k \left| \Omega_j^{(h)} \right| \right)^{1/2} \tag{6.63}$$

the global inequality

$$\left. \begin{array}{c} |\langle Bf, g \rangle| \\ |\langle Bg, f \rangle| \end{array} \right\} \le Ch^2 \, \|f; \, H^{-1/2}(\Gamma)\| \, \|g; \, L^2(\Omega)\|, \qquad \forall f \in S_h^1(\Gamma), \, g \in S_h^0(\Omega).$$

Collecting the above results, we obtain from (6.61), (6.62), (6.63) estimates for the admissable quadrature errors in order to ensure the global condition (6.44).

Lemma 6.11.1. *Let the mesh be quasi-uniform and let the cubature error for the calculation of each matrix entry be bounded by Ch^p whereby the exponent p is to be choosen as*

$$p := \left\{ \begin{array}{ll} 6, & \text{for } I - R^\Omega, \\ 5, & \text{for } V^\Gamma, \\ 11/2, & \text{for } R^\Gamma \text{ and } V^\Omega. \end{array} \right.$$

Then the condition (6.44) is fulfilled.

For graded meshes, the corresponding minimal and local values of the meshwidth must be inserted instead of h.

7. An Example: Katenoid Shell Under Torsion

This chapter illustrates the application of the BDIM to the special case of a katenoid[1] shell under pure torsion. In this case, the shell equations possess an analytical solution, the numerical solution can be compared to it.

We will study the Dirichlet problem, calculate the kernels of the integral operators and describe a strategy for its discretization. For the approximation error of the displacement field, measured in the L^2norm, we observed a good coincidence with the theoretical order $h^{2-\varepsilon}$, obtained in section 6.10.

7.1 Geometry of the Mid-Surface

In section 4.8 we have already explained, that the shell equations (4.48) have the simplest structure for an isothermal parametrization of the mid-surface. Besides circular cylindrical surfaces, the katenoids are the only axisymmetric surfaces for which cylindrical coordinates are already an isothermal parametrization. Circular cylinders lead to equations with constant coefficients. It was of interest however, to study the BDIM for equations with variable coefficients. That is why we choosed a katenoid shell as the simplest example fulfilling this requirement.

In cylindrical coordinates $x_1 = \psi$, $x_2 = \zeta$, the mid-surface possesses the parametrization

$$\mathbf{r} = R \begin{pmatrix} \cosh \zeta \, \cos \psi \\ \cosh \zeta \, \sin \psi \\ \zeta \end{pmatrix}$$

(see Fig. 7.1). R denotes the diameter of the body of revolution at $\zeta = 0$. The covariant base vectors are:

$$\mathbf{a}_1 = R \begin{pmatrix} -\cosh \zeta \, \sin \psi \\ \cosh \zeta \, \cos \psi \\ 0 \end{pmatrix}, \qquad \mathbf{a}_2 = R \begin{pmatrix} \sinh \zeta \, \cos \psi \\ \sinh \zeta \, \sin \psi \\ 1 \end{pmatrix},$$

$$\mathbf{a}_3 = \frac{1}{\cosh \zeta} \begin{pmatrix} \cos \psi \\ \sin \psi \\ -\sinh \zeta \end{pmatrix}.$$

---

[1] An axisymmetric surface with a cosh-profile.

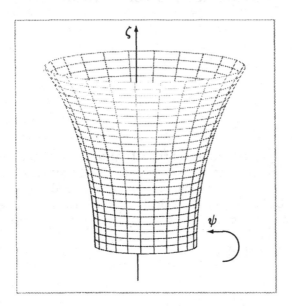

Fig. 7.1. A katenoid in cylindrical coordinates

This gives the first fundamental tensor

$$a_{11} = a_{22} = R^2 \cosh^2 \zeta, \qquad a_{12} = 0, \qquad a = R^4 \cosh^4 \zeta,$$

the second fundamental tensor

$$b_{11} = -R, \quad b_{22} = R, \quad b_{12} = 0, \quad b_1^1 = -\frac{1}{R \cosh^2 \zeta}, \quad b_2^2 = \frac{1}{R \cosh^2 \zeta}$$

and the Christoffel symbols

$$\Gamma_{11}^1 = \Gamma_{22}^1 = \Gamma_{12}^2 = 0, \qquad \Gamma_{12}^1 = \Gamma_{22}^2 = \tanh \zeta, \qquad \Gamma_{11}^2 = -\tanh \zeta$$

(cmp. (4.56)). The mean curvatur vanishes, related to the fact that the katenoid is a minimal surface. Gaussian curvature amounts to

$$\mathcal{K} = -\frac{1}{R^2 \cosh^4 \zeta}.$$

The components of the integration point y are denoted by φ and z:

$$x = (x_1, x_2) = (\psi, \zeta), \qquad y = (y_1, y_2) = (\varphi, z).$$

For the distance vector (5.10) and expression (5.15) we obtain the values:

$$\vartheta_1 := \psi - \varphi, \qquad \vartheta_2 := \zeta - z, \qquad \rho := \sqrt{\vartheta_1^2 + \vartheta_2^2}.$$

Contrary to the notation in chapter 5, the components of the distance vector are equipped now with lower indices. This change of the notation is performed since the indices can now be better distinguished from exponents.

7.2 Shell Equations and Boundary Conditions

Inserting the expressions from the previous section into the shell equations (4.48), we obtain the system

$$\sum_{k=1}^{3} A^{jk}(x)\, v_k(x) = -\sqrt{a}\, p^j(x), \qquad j = 1, 2, 3 \tag{7.1}$$

of differential equations with the following operators:

$$\begin{pmatrix} A^{11} & A^{12} \\ A^{21} & A^{22} \end{pmatrix}$$

$$= \frac{D}{R^2}\left[\begin{pmatrix} \partial_1 & -\partial_2 \\ \partial_2 & \partial_1 \end{pmatrix} \frac{1}{\cosh^2 \zeta} \begin{pmatrix} 1 & 0 \\ 0 & \nu \end{pmatrix} \begin{pmatrix} \partial_1 & \partial_2 \\ -\partial_2 & \partial_1 \end{pmatrix} - \frac{1-\nu}{\cosh^4 \zeta} I \right],$$

$$A^{33} = -\frac{B}{R^2}\left[\Delta \frac{1}{\cosh^2 \zeta} \Delta - (1-\nu) \begin{pmatrix} \partial_1 \\ \partial_2 \end{pmatrix}^{\mathsf{T}} \frac{1}{\cosh^4 \zeta} \begin{pmatrix} \partial_1 \\ \partial_2 \end{pmatrix} \right] - \frac{2\,D(1-\nu)}{\cosh^2 \zeta},$$

$$\begin{pmatrix} A^{13} \\ A^{23} \end{pmatrix} = \frac{D(1-\nu)}{R\cosh^2 \zeta} \begin{pmatrix} \partial_1 \\ -\partial_2 \end{pmatrix},$$

$$\begin{pmatrix} A^{31} \\ A^{32} \end{pmatrix}^{\mathsf{T}} = \frac{D(1-\nu)}{R} \begin{pmatrix} -\partial_1 \\ \partial_2 \end{pmatrix}^{\mathsf{T}} \frac{1}{\cosh^2 \zeta}.$$

Let the katenoid be bounded from below and above by the planes ζ_- and ζ_+, respectively. Now, the domain Ω in the parameter plane is the rectangle

$$\Omega = \left\{ (\psi, \zeta) \,\middle|\, : \psi \in (0, 2\pi);\ \zeta \in (\zeta_-, \zeta_+) \right\}. \tag{7.2}$$

The unit tangential vector $\mathbf{t}$ and the unit normal vector $\mathbf{n}$ have following components on the upper and lower edge of the mid-surface boundary:

$$\begin{pmatrix} t_1 \\ t_2 \end{pmatrix} = R\cosh \zeta_\pm \begin{pmatrix} \mp 1 \\ 0 \end{pmatrix}, \qquad \begin{pmatrix} n_1 \\ n_2 \end{pmatrix} = R\cosh \zeta_\pm \begin{pmatrix} 0 \\ \pm 1 \end{pmatrix}.$$

It can easily be verified that the displacement field

$$v_1 = \cosh \zeta \,\sinh(\zeta - \zeta_0), \qquad v_2 = v_3 = 0 \tag{7.3}$$

satisfies the system (7.1) for vanishing surface loads. Hereby, ζ_0 is an arbitrary constant. All strain components (4.6), (4.7) vanish, except

$$\alpha_{(12)} = \frac{1}{2} \cosh \zeta_0.$$

Thus, all stress components (4.9), (4.10) vanish, except

$$n^{(12)} = \frac{D(1 - \nu) \cosh \zeta_0}{2R^4 \cosh^4 \zeta}.$$

The tangential boundary traction on the upper and lower edge is equal to

$$\tau_n = \frac{D(1 - \nu) \cosh \zeta_0}{2R^2 \cosh^2 \zeta_\pm}.$$

The other boundary tractions and moments from (6.8) vanish. Thus the katenoid shell has the property, so that the outer form of the mid-surface is preserved under pure torsion. Only a membrane shearing occurs.

For our numerical tests with the Dirichlet problem, the boundary values of the displacement field (7.3) where taken as INPUT (with $\zeta_0 = 0$). The knowledge of the exact solution offers a good possibility to check the numerical solution. This concerns however, only the "final result", namely the displacement field. Unfortunately, the "intermediate results", i. e., the solution of the integral equation system, cannot be checked directly.

7.3 Levi Function and Integral Kernels

Here we collect the formulas for the kernels of all integral operators occurring in the representation formula (6.21) and in the integral equation system (6.31). Its detailed calculation, though only of importance for the implementation, is explained here in order to give an impression of this practical component of the method.

7.3.1 The Levi Function L^Ω

This matrix-function is given by (6.22). Due to (5.44) and (5.46), the individual entries are:

$$L^\square = \begin{pmatrix} L_{11} & L_{12} \\ L_{21} & L_{22} \end{pmatrix} =$$

$$= \frac{R^2 \cosh^2 z}{4\pi D(1 - \nu)} \left[(3 - \nu) I \ln \rho - \frac{1 + \nu}{\rho^2} \begin{pmatrix} \vartheta_1^2 & \vartheta_1 \vartheta_2 \\ \vartheta_1 \vartheta_2 & \vartheta_2^2 \end{pmatrix} \right],$$

$$L_{33} = -\frac{R^2 \cosh^2 z}{8\pi B} \rho^2 (\ln \rho - \gamma_0 + \tanh z \, \vartheta_2 \, \ln \rho),$$

$$L_{31} = L_{32} = L_{13} = L_{23} = 0.$$

The constant γ_0 was introduced, to have a simple possibility for a modification of the Levi function if an eigenvalue is catched by accident. The value of γ_0 can be choosen arbitrary. In the tests we set $\gamma_0 = 1/2$. An eigenvalue was never caught.

7.3.2 The Integral Kernel $L^\Gamma(x, y)$

From (6.23) we obtain:

$$\partial_1^\circ = \frac{\partial}{\partial\varphi}, \qquad \partial_2^\circ = \cosh^2(z)\, \frac{\partial}{\partial z}\, \cosh^{-2}(z)$$

$$\begin{pmatrix} \partial_1^\circ \\ \partial_2^\circ \end{pmatrix} L_{33} = c_3 \left\{ \left[\left(\ln\rho + \frac{1}{2} \right)(1 + \vartheta_2 \tanh z) - \gamma_0 \right] \begin{pmatrix} \vartheta_1 \\ \vartheta_2 \end{pmatrix} \right.$$
$$\left. + \frac{\rho^2 \ln\rho}{2} \left(\tanh z - \frac{\vartheta_2}{\cosh^2 z} \right) \begin{pmatrix} 0 \\ 1 \end{pmatrix} \right\}$$

7.3.3 Further Notations

$$c_0 = \frac{\cosh^2 z}{4\pi \cosh^2 \zeta}, \qquad k_1 = \ln\rho + 1/2 - \gamma_0,$$

$$k_3 = (\vartheta_1^2 + 3\vartheta_2^2) \ln\rho + \vartheta_2^2, \qquad k_4 = (1 + \tanh^2 z),$$

$$c_1 = c_0 \frac{6R(1 - \nu)}{d^2}, \qquad c_2 = \frac{R^2 \cosh^2 z}{4\pi D(1 - \nu)},$$

$$c_3 = \frac{R^2 \cosh^2 z}{4\pi B},$$

$$\sigma = \frac{3 - \nu - 4\sinh^2 \zeta}{2\cosh^2 \zeta}, \qquad \Theta = \frac{R}{d},$$

$$a_{33} = \frac{2\vartheta_2 \tanh\zeta}{\rho^2} + (\ln\rho + 1 - \gamma_0)\,\sigma - \frac{(1 - \nu)\tanh\zeta\, k_1\vartheta_2}{\cosh^2 \zeta}$$
$$- 3(1 - \nu)\Theta^2 \rho^2 (\ln\rho - \gamma_0),$$

$$r_{33} = -\frac{2\vartheta_2}{\rho^2} + 4\tanh\zeta \left(\ln\rho + \frac{3}{4} + \frac{\vartheta_2^2}{\rho^2} \right) + 2\sigma\vartheta_2 \left(\ln\rho + \frac{3}{4} \right)$$
$$- \frac{(1 - \nu)k_3 \tanh\zeta}{2\cosh^2 \zeta} - 3(1 - \nu)\Theta^2 \rho^2 \vartheta_2 \ln\rho.$$

Recall that d denotes the shell thickness (see Introduction).

7.3.4 The Integral Kernels V^Ω and V^Γ

The following expressions must be inserted into (6.26) and (6.27):

$$\begin{pmatrix} \frac{\partial}{\partial x_1} \\ \frac{\partial}{\partial x_2} \end{pmatrix} L_{33} =$$

$$= -c_3 \left\{ \left[k_1 + \vartheta_2 \tanh z \left(\ln \rho + \frac{1}{2} \right) \right] \begin{pmatrix} \vartheta_1 \\ \vartheta_2 \end{pmatrix} + \frac{1}{2} \tanh z \, \rho^2 \ln \rho \begin{pmatrix} 0 \\ 1 \end{pmatrix} \right\},$$

$$\begin{aligned} L_\square \;=\; c_3 \bigg\{ & k_1 \begin{pmatrix} 1 & 0 \\ 0 & 1 \end{pmatrix} + \frac{1 + \vartheta_2 \tanh z}{\rho^2} \begin{pmatrix} \vartheta_1^2 & \vartheta_1 \vartheta_2 \\ \vartheta_1 \vartheta_2 & \vartheta_2^2 \end{pmatrix} \\ & + \left(\ln \rho + \frac{1}{2} \right) \tanh z \begin{pmatrix} \vartheta_2 & \vartheta_1 \\ \vartheta_1 & 3\vartheta_2 \end{pmatrix} \\ & - \frac{1}{\cosh^2 z} \left[\left(\ln \rho + \frac{1}{2} \right) \vartheta_2 \begin{pmatrix} 0 & \vartheta_1 \\ 0 & \vartheta_2 \end{pmatrix} + \frac{\rho^2 \ln \rho}{2} \begin{pmatrix} 0 & 0 \\ 0 & 1 \end{pmatrix} \right] \bigg\}. \end{aligned}$$

7.3.5 The Integral Kernel R^Ω

For the components of the matrix

$$R^\Omega = \begin{pmatrix} R^{1\cdot}_{\cdot 1} & R^{1\cdot}_{\cdot 2} & R^{1\cdot}_{\cdot 3} \\ R^{2\cdot}_{\cdot 1} & R^{2\cdot}_{\cdot 2} & R^{2\cdot}_{\cdot 3} \\ R^{3\cdot}_{\cdot 1} & R^{3\cdot}_{\cdot 2} & R^{3\cdot}_{\cdot 3} \end{pmatrix}$$

we obtain from (5.75) the expressions:

$$\{R^{\alpha\cdot}_{\cdot\beta}\}^2_{\alpha,\beta=1} \;=\; c_0 \bigg\{ \frac{4 \tanh \zeta}{\rho^2} \begin{pmatrix} \vartheta_2 & -\vartheta_1 \\ \vartheta_1 & \vartheta_2 \end{pmatrix} +$$

$$+ \frac{1}{\cosh^2 \zeta} \left[(3 - \nu) \, I \, \ln \rho - \frac{1 + \nu}{\rho^2} \begin{pmatrix} \vartheta_1^2 & \vartheta_1 \vartheta_2 \\ \vartheta_1 \vartheta_2 & \vartheta_2^2 \end{pmatrix} \right] \bigg\},$$

$$\begin{pmatrix} R^{3\cdot}_{\cdot 1} \\ R^{3\cdot}_{\cdot 2} \end{pmatrix} \;=\; 2R \, c_0 \left[\frac{1 - \nu}{\rho^2} \begin{pmatrix} \vartheta_1 \\ -\vartheta_2 \end{pmatrix} + \right. \tag{7.4}$$

$$\left. + \frac{1 + \nu}{\rho^2} \left(\frac{\vartheta_1^2 - \vartheta_2^2}{\rho^2} - \vartheta_2 \tanh \zeta \right) \begin{pmatrix} \vartheta_1 \\ \vartheta_2 \end{pmatrix} + (3 - \nu) \tanh \zeta \begin{pmatrix} 0 \\ \ln \rho \end{pmatrix} \right],$$

$$\begin{pmatrix} R^{1\cdot}_{\cdot 3} \\ R^{2\cdot}_{\cdot 3} \end{pmatrix} \;=\; c_1 \left[2 \, k_1 \begin{pmatrix} \vartheta_1 \\ -\vartheta_2 \end{pmatrix} + \tanh z \begin{pmatrix} 2 \, \vartheta_1 \vartheta_2 \, (\ln \rho + 1/2) \\ -k_3 \end{pmatrix} \right],$$

$$R^{3\cdot}_{\cdot 3} \;=\; 4 \, c_0 \, (a_{33} + r_{33} \, \tanh z).$$

7.3.6 The Integral Kernel R^Γ

It is determined according to (6.25) and has the form

$$
R^\Gamma = \begin{pmatrix}
R^1_{\cdot 1} & R^1_{\cdot 2} & \partial^\circ_1 R^1_{\cdot 3} & \partial^\circ_2 R^1_{\cdot 3} \\
R^2_{\cdot 1} & R^2_{\cdot 2} & \partial^\circ_1 R^2_{\cdot 3} & \partial^\circ_2 R^2_{\cdot 3} \\
R^3_{\cdot 1} & R^3_{\cdot 2} & \partial^\circ_1 R^3_{\cdot 3} & \partial^\circ_2 R^3_{\cdot 3}.
\end{pmatrix}
$$

with

$$
\partial^\circ_1 R^1_{\cdot 3} = -2\,c_1 \left[\left(\ln\rho + \frac{1}{2} + \frac{\vartheta^2_1}{\rho^2} \right)(1 + \vartheta_2\tanh z) - \gamma_0 \right],
$$

$$
\partial^\circ_2 R^1_{\cdot 3} = -2\,c_1\,\vartheta_1 \left[\frac{\vartheta_2}{\rho^2} + \tanh z \left(\ln\rho + \frac{1}{2} + \frac{\vartheta^2_2}{\rho^2} \right) - \frac{\vartheta_2}{\cosh^2 z}(\ln\rho + 1/2) \right],
$$

$$
\partial^\circ_1 R^2_{\cdot 3} = 2c_1\,\vartheta_1 \left[\frac{\vartheta_2}{\rho^2} + \tanh z \left(\ln\rho + \frac{1}{2} + \frac{\vartheta^2_2}{\rho^2} \right) \right],
$$

$$
\partial^\circ_2 R^2_{\cdot 3} = c_1 \left[2\frac{\vartheta^2_2}{\rho^2} + 2k_1 + \vartheta_2\tanh z \left(6\ln\rho + 3 + \frac{2\vartheta^2_2}{\rho^2} \right) - \frac{k_3}{\cosh^2 z} \right],
$$

$$
\partial^\circ_1 R^3_{\cdot 3} = -4\,c_0 \left(\frac{\partial a_{33}}{\partial\vartheta_1} + \tanh z \frac{\partial r_{33}}{\partial\vartheta_1} \right),
$$

$$
\partial^\circ_2 R^3_{\cdot 3} = 4\,c_0 \left[-\frac{\partial a_{33}}{\partial\vartheta_2} + \frac{r_{33}}{\cosh^2 z} - \tanh z \frac{\partial r_{33}}{\partial\vartheta_2} \right],
$$

$$
\frac{\partial a_{33}}{\partial\vartheta_1} = -\frac{4\vartheta_1\vartheta_2\tanh\zeta}{\rho^4} + \frac{\sigma\vartheta_1}{\rho^2} - \frac{\vartheta_1\vartheta_2(1-\nu)\tanh\zeta}{\rho^2\cosh^2\zeta} - 6(1-\nu)\Theta^2\vartheta_1\,k_1,
$$

$$
\frac{\partial a_{33}}{\partial\vartheta_2} = 2\tanh\zeta\frac{\vartheta^2_1 - \vartheta^2_2}{\rho^4} + \frac{\sigma\vartheta_2}{\rho^2} - \frac{(1-\nu)\tanh\zeta}{\cosh^2\zeta}\left(k_1 + \frac{\vartheta^2_2}{\rho^2} \right)
$$
$$
-6(1-\nu)\Theta^2\vartheta_2\,k_1,
$$

$$
\frac{\partial r_{33}}{\partial\vartheta_1} = \frac{4\vartheta_1\vartheta_2}{\rho^4} + \frac{4\left(\vartheta^2_1 - \vartheta^2_2\right)\vartheta_1\tanh\zeta}{\rho^4} + \frac{2\sigma\vartheta_1\vartheta_2}{\rho^2}
$$
$$
-\vartheta_1\frac{(1-\nu)\tanh\zeta}{\cosh^2\zeta}\left(\ln\rho + \frac{1}{2} + \frac{\vartheta^2_2}{\rho^2} \right) - 6(1-\nu)\Theta^2\vartheta_1\vartheta_2(k_1 + \gamma_0)
$$

$$
\frac{\partial r_{33}}{\partial\vartheta_2} = -2\frac{\vartheta^2_1 - \vartheta^2_2}{\rho^4} + \frac{4\left(3\vartheta^2_1 + \vartheta^2_2\right)\vartheta_2\tanh\zeta}{\rho^4} + 2\sigma\left(\ln\rho + \frac{3}{4} + \frac{\vartheta^2_2}{\rho^2} \right)
$$
$$
-\frac{(1-\nu)\vartheta_2\tanh\zeta}{2\cosh^2\zeta}\left(6\ln\rho + 3 + \frac{2\vartheta^2_2}{\rho^2} \right) - 3(1-\nu)\Theta^2\,k_3.
$$

7.3.7 Mapping onto the Neumann Data

The kernels of these operators are obtained from the mapping (6.16) where v must be replaced by the columns of the matrices L^Ω and L^Γ. The application to L^Ω leads to weak singular kernels but the boundary operator contains Cauchy-singular and hypersingular parts. According to a well-known technique, these parts are regularized by partial integration with respect to the boundary parameter s.

The corresponding calculations are tedious and the representation of the final result, the formulas for the (partially integrated) integral kernels, require about 8 pages alone.

7.4 The Shell Equations as a Singular Perturbed Problem

The quotient $\Theta = R/d$ can take a magnitude of about 10^3 to 10^4 for realistic shells. The smoothest terms in the integral kernels are multiplied with this big parameter, what is the counterpart to the fact that the highest derivatives in the shell equations (7.1) are multiplied with a small parameter of the magnitude Θ^{-1}.

In this sense, the shell equations are a singular perturbed problem and it is possible to apply the well-known methods from this theory. If doing so, the solution of the differential equation is sought in form of a power series with respect to Θ^{-1}. The consecutive terms of the series are determined succesively, where in each step a boundary value problem (derived from the original problem) is to be solved. A detailed description and investigation of this method can be found in [48].

For numerical purposes however, this method seems not to be the best choice. Firstly, it will be very expensive to solve all the involved boundary value problems, since this must be done numerically. Secondly, a parameter of the magnitude 10^k in the integral equations or, respectively, a parameter of magnitude 10^{-k} in the differential equations still doesn't lead to the typical effects of a singular perturbed problem, provided that the number of valid digits in the computer arithmetic is essentially higher than k.

7.5 Corner Singularities and Mesh Refinement

The corner singularities are determined by solving the transcendental equation (6.39). In the actual case of a rectangle holds $\omega = \pi/2$ for all corners. Setting the first factor from the left side of (6.39) equal to zero, we obtain the following equation after the substitution $\lambda = (2i\alpha)/\pi$:

$$\frac{2\gamma}{\pi} = \pm\frac{\sin\alpha}{\alpha}. \tag{7.5}$$

From the second factor we obtain after the substitution $\lambda = (2i\beta)/\pi$:

$$\frac{2\gamma}{3\pi} = \pm \frac{\sin \beta}{\beta}. \tag{7.6}$$

These are the same equations as they are known from the biharmonic equation, whereby (7.5) corresponds to the interiour Dirichlet problem and (7.6) corresponds to the exteriour Dirichlet problem [85]. Since the integral equations don't contain any information, as to which of both problems is to be solved, both kinds of singular functions accur.

For $\gamma = 1$, equation (7.5) has the only solution $\alpha = \pm \pi/2$, which implies $\lambda = i$. The corresponding function is not singular but smooth. This agrees with the well-known fact that the interiour Dirichlet problem for the biharmonic equation has singularities only for interiour angles greater or equal to $\varphi_0 \approx 0.813\,\pi$ [85]. For $\gamma = 1$, Equation (7.6) possesses two solutions in the strip $0 < \Im m\,\lambda < 1$, namely:

$$\lambda = 0.5444837365\,i, \tag{7.7}$$
$$\lambda = 0.9085291898\,i.$$

For the upper left block of the operator V^Γ, the constant γ takes the value $(1 + \nu)/(3 - \nu)$ due to (6.33). Equations (7.5) and (7.6) correspond in this case to the Dirichlet problem for the two-dimensional equations of elasticity. Since Poisson's ratio ν always belongs to the interval $[0, 1/2]$, there holds $1/3 \leq \gamma \leq 3/5$. But then, equation (7.5) only has solutions from the domain $|\Im m\,\lambda| \geq 1$. The solutions of (7.6) with the smallest positive imaginary part of λ are:

$$\lambda = 0.6223242422\,i, \quad \text{for} \quad \gamma = 1/3,$$
$$\lambda = 0.5898951315\,i, \quad \text{for} \quad \gamma = 3/5.$$

The smallest among all actual exponents of the singular functions is given in (7.7). Further we mention that all zeros of the transcendental equation (6.39) in the strip $0 < \Im m\,\lambda < 1$ are simple zeros.

Due to (6.57), we have to choose the refinement parameter

$$q \geq \frac{1}{0.5444837365} = 1.836602148 \tag{7.8}$$

in order to achieve the maximal order of convergence.

7.6 Discretization of the Integral Operators

We have to deal with four different types of integral operators, namely mappings from Ω to Ω, from Ω to Γ, from Γ to Ω and from Γ to Γ. Its kernels are weak singular or Cauchy-singular, dependent on the respective minimal degree of homogeneity. The following principles where applied for the discretization of all these operators.

The rectangle (7.2) is subdivided by horizontal and vertical lines into $M_1 \times M_2$ elements. The boundary is subdivided by these lines into $L := 2(M_1 + M_2)$ elements. The distances of the lines are refined to the q th power in the vicinity of each corner point, as described in section 6.9. The required value of the refinement parameter q is given by (7.8), but in order to study its influence, q was realized in the test version of our code as variable INPUT.

The set of trial functions for the Galerkin's method consists of the characteristic functions of the sub-rectangles and the hat functions on Γ, continuous even at the corner points. The hat functions are split into half-hats and those are integrated separately. (Note that in the case of hypersingular kernels such a separete integration of the half-hats is not directly applicable.)

For each entry of the discretization matrix, i. e., for each pair of trial functions, we have to evaluate 1, 2 or 4 integrals over a two-, three- or four-dimensional parallelepipedon. Hereby one has firstly to distinguish between the two cases: "near-field" or "far-field". If the supports of both actual trial functions have a positive distance, the case is handled as far-field, else as near-field.

7.6.1 Far-Field

In the far-field, real-analytic functions are to be integrated over parallelepipedons. We have used the Gaussian product cubature rule with the same number N of quadrature sample points for each direction of integration. (Note, that in more than one dimension the terminus *quadrature* usually is replaced by *cubature*, but we will us both terminologies synonymously.) In order to save computation time, the number N is apriori determined (for each actual multiple integral separtely) so that the local error estimations (6.61), (6.62) or (6.63), respectively, are fulfilled (see sect. 7.7).

7.6.2 Near-Field

The discretization of the near-field is the most complicated part in the development of the code. While the far field consumes the main part of computation time, the near field consumes the main part of the code developing time.

In the near field one has to distinguish, whether the supports are coincident or neighbouring. The later case means that the supports touch each other at a point or at an edge.

In the case of neighbouring supports, the integrand contains a singularity,[2] which is located at a corner, an edge or at an outer diagonal of the corre-

---

[2] "Singularity" means the set of all points where the integrand is not smooth.

sponding parallelepipedon. The domain of integration is split into subdomains, which are right-angled tringles in the two-dimensional case, prisms and pyramids in the higher-dimensional case. The splitting is done in such a way that the singularities can be smoothed by the introduction of "Duffy coordinates" [30], well-adapted to each subdomain. By such a substitution, each weak singular integrand is transformed into the sum of a smooth function and an other smooth function, multiplied by a logarithmic term. Thereby a logarithmic dependence occurs only with respect to one of the new variables.

The new domain of integration is a parallelepipedon again. The Gaussian product cubature can be applied to the smooth part. Here 2 or 3 sample points suffice. For the evaluation of the logarithmic part one applies a combined quadrature rule which consists on a weighted Gaussian quadrature rule according to [1] in the "logarithmic" direction and the usual cubature rule in the other directions.

More detailed descriptions of these splitting methods and the application of Duffy coordinates can be found e. g. in [54, sect. 3], [75] and [112].

The case of coincident supports occurs only for the operators V^Γ and R^Ω. The operator V^Γ is treated as follows. Suppose, the domain of integration is the square

$$(s, t) \in (0, h) \times (0, h)$$

and the (weak) singularity of the integrand is located on the diagonal $s = t$. The substitution

$$\xi = \frac{s - t}{h}, \qquad \eta = \frac{s + t - h}{h - |s - t|}, \qquad \frac{D(s, t)}{D(\xi, \eta)} = \frac{h^2}{2}(1 - |\xi|)$$

transforms the domain of integration into the square $(-1, 1) \times (-1, 1)$. The singularity is located now on the segment $\xi = 0; \eta \in (-1, 1)$. This segment divides the square into two rectangles and in the interiour of both, the integrand is infinitely smooth. In η-direction, a Gaussian quadrature rule with few (1 to 3) sample points can be applied. For the quadrature in ξ-direction, several possibilities come into question:

– a Gaussian quadrature with a high number of sample points;
– separation and analytical evaluation of the main singularity; quadrature of the remaining part with fewer sample points;
– separate quadrature of the logarithmic and the ln-free terms; application of the weighted Gaussian rule [1] to the logarithmic part.

The last method can be applied if the integrand has the minimal degree 0 of pseudohomogeneity. It was realized in our code for the discretization of V^Γ. The separate treatement of the logarithmic and the ln-free terms leads to the additional difficulty, so that the subroutines for the calculation of the kernel functions must supply the corresponding parts separately. If these subroutines are generated by computer algebra (what is meaningful in general), one has to pay attention to this distinction already in the algebraic procedures.

The operator $R^\Omega(x,y)$ was treated as follows in the case of coincident supports. The terms of degree -1 are separated from the kernel $R^\Omega(x,y)$ and integrated analytically. Gaussian product cubature with a higher number of sample points was applied to the remaining part. An other possibility would have been to carry over the substitution technique described above to the four-dimensional case.

7.7 Estimations for the Number of Cubature Sample Points

The question, how many quadrature sample points are necessary to achieve a pre-determined accuracy, is widely studied in several papers. The estimates given by Rabinowitz and Richter [99] for one-dimensional integrations are based on a continuation of the integrand to the complex plane. After transforming the integration to the interval $[-1,1]$, one consideres the family of all confocal ellipses $\mathcal{E}_\rho \subset \mathbb{C}$ with foci at ± 1, where $\rho = a + b$, a is the semimajor axis of $\mathcal{E}_\rho$ and $b = \sqrt{a^2 - 1}$ is the semiminor axis. If the integrand $f(z)$ is analytic in $\mathcal{E}_\rho$, the error of the N-point Gauss quadrature rule can be estimated by [99, formula (22)]:

$$|E(f)| \leq \frac{C(N,\rho)}{\rho^{2N}} \sup_{z \in \mathcal{E}_\rho} |f(z)| \tag{7.9}$$

with

$$C(N,\rho) \approx \pi \left[1 + \frac{1}{2\rho^4} \left(1 + \frac{1}{2N} \right) \right] \leq \frac{17}{8} \pi.$$

Note that no derivatives of the integrand enter into the estimation; but the location of the "next singularity" in the complex plane and the behaviour in the vicinity of this singularity are of importance. All integral kernels occuring within the BDIM are pseudohomogeneous with respect to $|x - y|$ and hence, they are analytic outside the singularity and its behaviour near the singularity is known. Thus the estimate (7.9) can be applied in a straightforward manner to determine the number N in such a way that a pre-determined accuracy is guaranteed. We will outline such estimates for the special case of the boundary integral operator V^Γ here since this example is also of interest for the BEM.

Let $K(x,y)$ be one of the kernels from the matrix $V^\Gamma(x,y)$. It is pseudo-homogeneous in $\vartheta = x - y$ of degree 0 and thus it admits an estimate

$$|K(x,y)| \leq C_1 + C_2 |\ln |x - y||, \quad \forall x,y \in \Gamma.$$

Let $\tilde{\chi}_k^\Gamma(x), \tilde{\chi}_j^\Gamma(y)$ be two half-hat functions with supports on the boundary segments $\Gamma_k^{(h)}, \Gamma_j^{(h)}$, respectively. Denote the midpoint of $\Gamma_j^{(h)}$ by $x^{(j)}$ and the unit tangential vector by $t^{(j)}$. The boundary segment $\Gamma_j^{(h)}$ has the length h_j and possesses the parametrization

$$\Gamma_j^{(h)} = \left\{ x^{(j)} + z\frac{h_j}{2}t^{(j)} \Big|: \ z \in [-1,1] \right\}. \tag{7.10}$$

Here we have used the fact that the boundary consists of straight lines but the following considerations can be applied to a curvilinear boundary as well. The representation (7.10) is to be understood then as an approximation.

We have to evaluate the double integral

$$a_{jk} := \int_\Gamma \int_\Gamma K(x,y)\, \tilde{\chi}_k^\Gamma(x)\, \tilde{\chi}_j^\Gamma(y) d\Gamma_x d\Gamma_y \tag{7.11}$$

$$= \frac{h_j h_k}{4} \int_{-1}^{1} \int_{-1}^{1} K[x(\zeta),y(z)]\, \tilde{\chi}_k^\Gamma[x(\zeta)]\, \tilde{\chi}_j^\Gamma[y(z)]\, d\zeta dz$$

$$= \frac{h_j h_k}{4} \int_{-1}^{1} K_1[y(z)]dz \tag{7.12}$$

with

$$K_1(y(z)) := \tilde{\chi}_j^\Gamma(y) \int_{-1}^{1} \tilde{\chi}_k^\Gamma[x(\zeta)]K[x(\zeta),y(z)]d\zeta \tag{7.13}$$

$$x(\zeta) = x^{(k)} + \zeta\frac{h_k}{2}t^{(k)},$$

$$y(z) = x^{(j)} + z\frac{h_j}{2}t^{(j)}.$$

Let $v, w \in \mathbb{R}^n$ be two arbitrary vectors with $w \neq 0$. The quadratic equation $|v - zw|^2 = 0$ has the pair of complex roots

$$z_\pm = \frac{|v|}{|w|}e^{\pm i\alpha}$$

where α is the angle between v and w. If we investigate the integration with respect to z and consider ζ as fixed, we have

$$x(\zeta) - y(z) = v - zw,$$

$$v = x^{(k)} + \zeta\frac{h_k}{2}t^{(k)} - x^{(j)}, \qquad w = \frac{h_j}{2}t^{(j)},$$

$$|z_\pm| = \frac{|v|}{|w|} \geq \frac{2D_{jk}}{h_j},$$

$$D_{jk} := \text{dist}\left(\Gamma_k^{(h)}, x^{(j)}\right).$$

Hence the integrand is analytic in the circle

$$|z| < \rho_0 := \frac{2D_{jk}}{h_j}. \tag{7.14}$$

Now we consider a smaller circle with radius $\rho_\varepsilon := (1-\varepsilon)\rho_0$. For $x = x(\zeta)$, $y = y(z)$ we have the representation

$$|x - y|^2 = |w|^2 (z - z_+)(z - z_-).$$

Assume that $|z| \leq \rho_\varepsilon$, $\Im z \geq 0$, $\Im z_+ \geq 0$. Then

$$|x - y|^2 \geq |w|^2 |z - z_+|^2 \geq |w|^2 \varepsilon^2 |z_+|^2 \geq \varepsilon^2 D_{jk}^2.$$

Thus we obtain the estimate

$$\sup_{|z| < \rho_\varepsilon} |K(x, y)| \leq C_1' + C_2' |\ln (\varepsilon D_{jk})|$$

with new constants C_1', C_2'. If ε is assumed to be bounded from below, the last expression is bounded from above by a certain constant C_3.

Analogously one can show that the function (7.13) is analytic and bounded in ρ_ε. If the integral (7.12) is replaced by an N-point Gaussian product cubature, we obtain from (7.9) an estimate of the quadrature error in the form

$$|a_{jk} - \tilde{a}_{jk}| \leq \frac{h_k h_j}{4} \cdot \frac{C C_3}{\rho_\varepsilon^{2N}}. \tag{7.15}$$

This estimate was derived under the assumptions that the integration with respect to x is performed exactly and the integration with respect to y is replaced by a N-point Gaussian quadrature. If the roles of x and y are changed, a similar estimate holds for the quadrature error, whereby in (7.14) D_{jk} is to be replaced by D_{kj}.

It can easily be verified (see e. g. [75, 112]) that the cubature error of a product formula can be estimated by the maximal quadrature error with respect to the single variables, multiplied by a constant.

Thus we obtain for the N-point Gaussian product cubature of the double integral (7.11) an error estimate of the form (7.15) with

$$\rho_\varepsilon = (1-\varepsilon)\rho_0, \quad \rho_0 = 2\max \left\{ \frac{\text{dist}\left(\Gamma_k^{(h)}, x^{(j)}\right)}{h_j}, \frac{\text{dist}\left(\Gamma_j^{(h)}, x^{(k)}\right)}{h_k} \right\}. \tag{7.16}$$

Inserting (7.15) into (6.62) gives an inequality which allows an estimation of the number N from below. In the case of a quasi-uniform mesh, the estimate takes the simple form

$$\frac{1}{\rho_\varepsilon^{2N}} \leq C h^3$$

with a certain constant C. This is equivalent to

$$N \geq \frac{-\ln C - 3\ln h}{2\ln \rho_\varepsilon}.$$

Similar estimates can be obtained for the other integral operators. In order to bring these results into a unified formulation, we assume that $\tilde{\chi}_k$ and $\tilde{\chi}_j$ are either half-hats or characteristic functions on the subdomains. We define

$$T_{jk} := 2\frac{\text{dist}\,(\text{supp}\,\tilde{\chi}_k,\,\text{supp}\,\tilde{\chi}_j)}{\max\,(\text{diam}\,\text{supp}\,\tilde{\chi}_k,\,\text{diam}\,\text{supp}\,\chi_j)}. \qquad (7.17)$$

Note that this number is always larger than 1, provided that the mesh fulfills standard requirements. Further, the number ρ_ε from (7.16) can be estimated from below by T_{jk}.

Lemma 7.7.1. *Assume that all multiple integrals for calculating the entries of the discretization matrix in the far-field are replaced by a N-point Gaussian product cubature. In case of a quasi-uniform mesh, the accuracy requirements* (6.61), (6.62), (6.63) *are fulfilled if we choose*

$$N \geq \frac{C_Q + \kappa|\ln(h/l)|}{\ln T_{jk}}, \qquad (7.18)$$

whereby T_{jk} is defined by (7.17), C_Q is a certain constant, l is a unit length and

$$\kappa := \begin{cases} 3/2, & \text{for } V^\Gamma, \\ 5/4, & \text{for } R^\Gamma, V^\Omega, \\ 1, & \text{for } R^\Omega. \end{cases}$$

We want to add some remarks[3]:

1. For the unit lenght l one can choose e. g. the diameter of the domain Ω. In our tests we have used $l = 1$.
2. For graded meshes, estimations of N can be derived by the same principles; only the different values of h come into consideration.
3. The above estimates for N are slightly rough. Better estimates can be obtained by considering the ellipses $\mathcal{E}_\rho$ instead of the circles. This requires the evaluation of the actual geometrical situation in each concrete situation. This evaluation costs about 15 commands in the code (without loops) and brings advantages mainly in the "near far-field". Note that each calculation of a kernel function usually costs essentially more, since the kernels in the BDIM are represented by lengthy formulas.
4. We are here in the situation of simple geometry. All domains of integration are parallelepipedons. In the general case one has a triangulation of the domain. But the necessary number of cubature sample points can be estimated as well. Therefore we refer to the techniques described in [109].
5. The number C_Q in (7.18) is still undetermined and we can choose different constants for the different operators. In practice these constants can be determined immediately by numerical experiments. The realization is

---

[3] Remark 6 is due to O. Steinbach, who confirmed it by numerical experiments with the BEM (private communication).

quite easy: one increases C_Q step by step till the numerical solution shows no further improvement. But this strategy is only applicable if one considers the operator with one fixed right-hand side (this was the case in our numerical experiments).

6. If one wants to ensure a good behaviour of the disturbed Galerkin method (6.45) for all possible right-hand sides, the determination of the constant C_Q is very non-trivial. The choosing of N according to the above estimates ensures that both summands in the square bracket on the right-hand side of (6.47) have the same asymptotic behaviour for $h \to 0$, namely h^2. But the second summand depends only on the operator and not on the solution $\mathbf{f}$. If the norm of the solution $\mathbf{f}$ is small, the second summand can produce an arbitrary large relative error if the cubature is not exact enough. To overcome this difficulty one has to choose the constant C_K in (6.44) and all constants derived from it in dependence of the solution, which is apriori unknown. This problem can be considered as unsolved at the present stage.

7.8 Numerical Results

We have implemented the indirect method for the katenoid shell in order to test the BDIM. For the parameter values

E (= Young's modulus) = 3000,
ν (= Poisson's ratio) = 0.2,
d (= shell thickness) = 0.3,
R (= minimal diameter of the body of revolution) = 30,
$\zeta_0 = \zeta_- = 0, \quad \zeta_+ = 1,$

we obtained the following results:

N	M	h	Dim	ϵ	ϵ/h^2	t_W
6	2	0.910	101	0.06	0.072	2
9	3	0.607	178	0.017	0.046	7
12	4	0.455	273	0.0086	0.042	13
15	5	0.364	386	0.0062	0.047	26

The notations in the table are:

N = Number of subintervals in ψ direction;

M = Number of subintervals in ζ direction;

h = Mean value of the meshwidth (= circumference of the rectangle divided by L);

Dim = dimension of the discretization matrix;

ε = error of the displacement field in the norm $L^2(\Omega)$;

t_W = time on a workstation with a MIPS R4000-processor (100 MHz) in seconds.

The linear systems where solved by the classical Gaussian algorithm. It consumed about 40% of the computation time on the finest mesh, thus one can still save some time here by using iterative solvers.

The results of the above table are achieved on a graded mesh with the refinement parameter (7.8). As expected, no improvement of the order of convergence (and also no reduction of the errors) was observed for stronger graduations. On an uniform mesh(for $q = 1$), the L^2-error of the displacement field was nearly proportional to h.

The constant C_Q from (7.18) appeared to be best choosen for, from 1 to 1.5. This leads to a mean number of quadrature sample points of about 2 to 2.5 in each direction of integration. In the "far far-field" one (!) quadrature sample point often suffices to ensure the necessary accuracy. In the "near far-field", the number of quadrature sample points reached 7 at the most. Calculations with an enlarged number of C_Q did not lead to more accurate results, though the integrations were performed more exactly. This can be explained by the other approximation errors.

A calculation[4] of the same example with FEM consumed more computation time (about 35 seconds).

---

[4] The FEM-package PSU was used herefore. Dipl.-Ing. F. Buschbeck is acknowledged for performing this calculations.

Index

References

1. D. G. Anderson. Gaussian quadrature formulae for $\int -\ln(x)f(x)dx$. *Math. Comp.* **19**, 477–481, 1965.
2. H. Antes. Über singuläre Lastfälle in einer linearen Schalentheorie. *Ingenieur-Archiv* **45**, 99–114, 1976.
3. H. Antes. On boundary integral equations for circular cylindrical shells. In *Proc. 3rd Int. Seminar*, Irvine, USA, July 1981.
4. T. Apel, A. M. Sändig and J. R. Whiteman. Graded mesh refinement and error estimates for finite element solutions of elliptic boundary value problems in non-smooth domains. *Math. Meth. Appl. Sci.* **19**, 63–85, 1996.
5. D. N. Arnold and W. Wendland. On the asymptotic convergence of collocation methods. *Math. Comput.*, **41**, 349–381, 1983.
6. J. P. Aubin. *Approximation of Elliptic Boundary Value Problems*. Wiley-Interscience, New York et. al., 1972.
7. A. Avantaggiati and M. Troisi. Spazi di Sobolev con peso e problemi ellitici in un angolo, I, II, III. *Ann. Mat. Pura Appl.* **95**, 361–408, 1973; **97**, 207–252, 1973; **99**, 1–51, 1974.
8. I. Babuška and C. Schwab. A-posteriori error estimation for hierarchic models of elliptic boundary value problems on thin domains. *SIAM J. Num. Anal.* **33**, 221–246, 1996.
9. Y. Başar and W. B. Krätzig. *Mechanik der Flächentragwerke*. Vieweg, Braunschweig, 1985.
10. M. Bernadou, P. G. Ciarlet and B. Miara. Existence theorems for two-dimensional linear shell theories. *Journal of Elasticity* **34**, 111–138, 1994.
11. D. E. Beskos. Dynamic analysis of plates and shallow shells by the D/BEM. In G. Z. Voyiadis and D. Karamanlidis, editors, *Advances in the theory of plates and shells*, pages 177–196, Amsterdam, 1990.
12. D. E. Beskos. Static and dynamic analysis of shells. In D. E. Beskos, editor, *Boundary element analysis of plates and shells*, Springer, Berlin, pages 93–140, 1991.
13. C. A. Brebbia, J. C. F. Telles and L. C. Wrobel *Boundary element techniques*. Springer, Berlin, 1984.
14. M. Bourlard, S. Nicaise and L. Paquet. An adapted boundary element method for the Dirichlet problem in polygonal domains. *SIAM J. Num. Anal.* **28**, 728–743, 1991.
15. K. F. Chernykh. Relation between dislocations and concentrated loadings in the theory of shells. *J. Appl. Math. Mech. (PMM)* **23**, 359–371, 1959.
16. G. N. Chernyshev. On the action of concentrated forces and moments on elastic thin shells of arbitrary shape. *J. Appl. Math. Mech. (PMM)* **27**, 172–184, 1963.
17. G. N. Chernyshev. Representation of solutions of the Green type equations of shells by the small parameter method. *J. Appl. Math. Mech. (PMM)* **32**, 1083–1089, 1968.

18. P. G. Ciarlet. *The Finite Element Method for Elliptic Problems.* North-Holland Pub. Comp., Amsterdam et. al., 1978.

19. D. L. Clements. A boundary integral equation method for the numerical solution of a second order elliptic equation with variable coefficients. *J. Austral. Math. Soc.* **22**, 218–228, 1980.

20. D. L. Clements. The boundary element method for linear elliptic equations with variable coefficients. In C. Brebbia, editor, *Boundary Elements X*, pages 91–96, 1988.

21. D. Colton and W. Wendland. Constructive methods to solve the exterior Neumann problem for the reduced wave equation in a spherically symmetric medium. *Proc. of the Royal Soc. Edinburgh* **A 75**, 97–107, 1976.

22. M. Costabel. Boundary integral operators on curved polygons. *Ann. Mat. Pura Appl.* **133**, 305–326, 1983.

23. M. Costabel. Boundary integral operators on Lipschitz domains: elementary results. *SIAM J. Math. Anal.* **19**, 613–626, 1988.

24. M. Costabel and E. Stephan. Boundary integral equations for mixed boundary value problems in polygonal domains and Galerkin approximation. *Banach Center Publications* **15** PWN, 175–251, 1985.

25. M. Costabel and W. L. Wendland. Strong ellipticity of boundary integral operators. *Crelle's Journal für die Reine und Angew. Math.* **372**, 34–63, 1986.

26. M. Costabel, E. Stephan and W. L. Wendland. Zur Randintegralmethode für das erste Fundamentalproblem der ebenen Elastizitätstheorie auf Polygongebieten. In H. Kurke ed. al., editor, *Recent trents im mathematics*, volume 50, pages 56–68. Teubner-Texte zur Mathematik, Leipzig, 1982.

27. M. Costabel, E. Stephan and W. L. Wendland. On boundary integral equations of the first kind for bi-Laplacians in a polygonal plane domain. *Ann. Sc. Norm. Sup., Pisa* **X**, Serie IV, 197–241, 1983.

28. R. Doré and W. Flügge. Force singularities of shallow cylindrical shells. *J. Appl. Mech., Transactions of ASME* **37**, 361–366, 1970.

29. A. Douglis and L. Nirenberg. Interior estimates for elliptic systems of partial differential equations. *Commun. Pure Appl. Math.* **8**, 503–538, 1955.

30. M. G. Duffy. Quadrature over pyramid or cube of integrands with a singularity at a vertex. *SIAM J. Num. Anal.* **19**, 1260–1262, 1982.

31. N. W. Efimow. *Flächenverbiegung im Grossen.* Akademie-Verlag, Berlin, 1957.

32. R. E. Elling. Concentrated loads applied to shallow shells. *J. Engng. Mech. Div. ASCE* **99**, 319–330, 1973.

33. J. Elschner. The double layer potential operator over polyhedral domains II: spline Galerkin methods. *Math. Meth. Appl. Sci.* **15**, 23–37, 1992.

34. J. Elschner. *Singular ordinary differential operators and pseudodifferential equations.* Akademie-Verlag, Berlin, 1985.

35. G. I. Èskin. *Boundary value problems for elliptic pseudodifferential equations.* American Math. Soc., Providence, Rhode Island, 1981.

36. M. Feistauer. On the finite element approximation of functions with noninteger derivatives. *Numer. Funct. Anal. and Optimiz.* **10**, 91–110, 1989.

37. G. Fichera. *Una introduzione alla teoria delle equazioni integrali sigolari.* Edizione Cremonese, Roma, 1958.

38. G. Fichera. Linear elliptic equations of higher order in two independent variables and singular integral equations, with applications to anisotropic inhomogeneous elasticity. In *Proc. Conference on Partial Differential Equations and Continuum Mechanics*, pages 55–80, Madison, 1961. Univ. of Wisconsin Press.

39. G. Fichera and P. Ricci. *The single layer potential approach of boundary value problems for elliptic equations.* Lecture Notes in Mathematics **561**, Springer, Berlin, pages 39–50, 1976.

40. W. Flügge. Concentrated forces on shells. In H. Görtler, editor, *Proc. 11th Int. Cong. Appl. Mech.*, pages 270–276, 1964.

41. W. Flügge and D. A. Conrad. Thermal singularities for cylindrical shells. In *Proc. 3rd U.S. Nat. Congr. Appl. Mech.*, pages 321–328, Berlin, 1958.

42. W. Flügge and R. E. Elling. Singular solutions for shallow shells. *Int. J. Solides Structures* **8**, 227–247, 1972.

43. D. J. Forbes and A. R. Robinson. Numerical analysis of elastic plates and shallow shells by an integral equation method. Ad691275, Univ. of Illinois, Illinois, 1969. ASTIA Doc.

44. K. Foresberg and W. Flügge. Point load on a shallow elliptic paraboloid. *J. Appl. Mech., Transactions of ASME* **33**, 575–585, 1966.

45. M. Galler. Fundamentallösungen von homogenen Differentialoperatoren. *Dissertationes Mathematicae* **282**, 1989.

46. I. M. Gel'fand and G. E. Shilov. *Generalized Functions I.* Academic Press, New York, London, 1964.

47. A. L. Gol'denveizer. Question of analyzing shells for concentrated loads (in russian). *J. Appl. Math. Mech. (PMM)* **18**, 181–186, 1954.

48. A. L. Gol'denveizer. Some mathematical problems in the linear theory of elastic thin shells (in russian). *Uspechi Mat. Nauk* **15**, 3–75, 1960.

49. G. Gospodinov. The boundary element method applied to shallow spherical shells. In Brebbia, editor, *Proc. 6th Int. Conf. Boundary Elements*, pages 3.65–3.75, 1984.

50. P. Grisvard. *Elliptic problems in nonsmooth domains.* Pitman, Boston et. al., 1985.

51. O. v. Grudzinski. *Quasihomogeneous Distributions.* North-Holland, Amsterdam et. al., 1991.

52. W. Haack and W. Wendland. *Vorlesungen über partielle und Pfaffsche Differentialgleichungen.* Birkhäuser Verlag, Basel, Stuttgart, 1969.

53. W. Hackbusch. *Integralgleichungen. Theorie und Numerik.* Teubner, Stuttgart, 1989. Engl. translation: *Integral equations.* To be published by Birkhäuser, Basel.

54. W. Hackbusch, C. Lage and S. A. Sauter. Efficient realization of the BEM. In W. L. Wendland, editor, *Boundary Element Topics*, Springer, Berlin, to appear 1997.

55. J. Hadamard. *Lectures on Cauchy's problem in linear partial differential equations.* Dover Publications, New York, 1952.

56. L. M. Hadjikov, S. Marginov and P. T. Bekyarova. Cubic spline boundary element method for circular cylindrical shells. In C. A. Brebbia, editor, *Proc. 7th Int. Conf. Boundary Elements*, pages 4.93 – 4.102, 1985.

57. F. Hartmann. *Methode der Randelemente.* Springer, Berlin, 1987.

58. J.-C. Hein. *Eine Integralgleichungsmethode zur Lösung eines gemischten Randwertproblemes der beliebig belasteten Kreiszylinderschale mit Ausschnitten glatter Berandung.* PhD thesis, TH Darmstadt, 1979.

59. D. Hilbert. *Jahresbericht der deutschen Mathematiker-Vereinigung* **16**, 77–78, 1907.

60. D. Hilbert. *Grundzüge einer allgemeinen Theorie der linearen Integralgleichungen.* Teubner, Leipzig, 1912.

61. S. Hildebrandt and E. Wienholtz. Constructive proofs of representation theorems in separable hilbert space. *Commun. Pure Appl. Math.* **17**, 369–373, 1964.

62. L. Hörmander. *The analysis of linear partial differential operators*, volume I. Springer, Berlin, 1985.

63. L. Hörmander. *The analysis of linear partial differential operators*, volume III. Springer, Berlin, 1985.

64. G. Hsiao and R. C. MacCamy. Solution of boundary value problems by integral equations of the first kind. *SIAM Review* 15, 687–704, 1973.

65. G. C. Hsiao, P. Kopp and W. L. Wendland. A Galerkin collocation method for some integral equations of the first kind. *Computing* 25, 89–130, 1980.

66. G. C. Hsiao, P. Kopp and W. L. Wendland. Some applications of a Galerkin-collocation method for boundary integral equations of the first kind. *Math. Meth. Appl. Sci.* 6, 280–325, 1984.

67. G. C. Hsiao and W. L. Wendland. A finite element method for some integral equations of the first kind. *J. Math. Anal. Appl.* 58, 449–481, 1977.

68. G. C. Hsiao and W. L. Wendland. The Aubin-Nitsche lemma for integral equations. *Journ. of Integral Equations* 3, 299–315, 1981.

69. G. C. Hsiao and W. L. Wendland. *Variational methods for boundary integral equations*. Springer, Berlin, (in preparation).

70. A. Jahanshahi. Force singularities of shallow cylindrical shells. *J. Appl. Mech., Transactions of ASME* 30, 342–346, 1963.

71. A. Jahanshahi. Some notes on singular solutions and the Green's function in the theory of plates and shells. *J. Appl. Mech., Transactions of ASME* 31, 441–446, 1964.

72. F. John. The fundamental solution of linear elliptic differential equations with analytic coefficients. *Commun. Pure Appl. Math.* 3, 273–304, 1950.

73. F. John. *Plane waves and spherical means applied to partial differential equations*. Interscience Publisher, New York, 1955.

74. R. Kieser. *Über einseitige Sprungrelationen und hypersinguläre Operatoren in der Methode der Randelemente*. PhD thesis, University of Stuttgart, 1991.

75. R. Kieser, C. Schwab, and W. L. Wendland. Numerical evaluation of singular and finite-part integrals on curved surfaces using symbolic manipulation. *Computing* 49, 279–301, 1992.

76. W. T. Koiter. A spherical shell under point loads at its poles. In *Progress in Applied Mechanics - The Prager Anniv. Vol.*, pages 155–169, 1963.

77. V. A. Kontratiev. Boundary problems for elliptic equations in domains with conical or angular points. *Trans. Moscow Math. Soc.* 16, 227–313, 1967.

78. Ho Kwang-Chien and Chen Fu. Simplified method of analysis for elliptic paraboloidal shallow shells under the action of concentrated loads. *Int. Assoc. Bridge Struc. Eng.* 24, 125–141, 1964.

79. M. V. Lazarenko and V. I. Tarakanov. On the fundamental solution of the equations of cylindrical shells (in russian). In *Mechanics of Continual Media*, pages 35–47. Tomsk, 1983.

80. X. Y. Lei and M. G. Huang. Boundary element method for shallow spherical shell bending problems involving shear deformation. In C. A. Brebbia et. al., editor, *Boundary Element Methods in Engineering, 9th Int. Conf. on BEM*, pages 69–80, 1987.

81. E. E. Levi. I problemi dei valori al contorno per le equazioni lineari totalmente ellittiche alle derivate parziali. *Mem. Soc. It. dei Sc. XL* 16, 1909.

82. Ju. I. Ljubič. On the existence "in the large" of fundamental solutions of second order elliptic equations (in russian). *Math. Sbornik* 57, Nr. 99, 45–58, 1962.

83. S. Lukasiewicz. Concentrated loads on shallow spherical shells. *Quarterly. J. Mech. Appl. Math.* 20, 293–305, 1967.

84. S. Lukasiewicz. The solution of concentrated loads on shells by means of Thomson functions. *ZAMM* 48, 247–254, 1968.

85. H. Melzer and R. Rannacher. Spannungskonzentrationen in Eckpunkten der Kirchhoffschen Platte. *Der Bauingenieur* **55**, 181–184, 1980.
86. S. G. Michlin. *Partielle Differentialgleichungen in der mathematischen Physik.* Akademie-Verlag, Berlin, 1978.
87. S. G. Mikhlin. *Multidimensional singular integrals and integral equations.* Pergamon Press, Oxford et. al, 1965.
88. S. G. Mikhlin, editor. *Linear equations of mathematical physics.* Holt, Rinehard and Winston, New York, 1967.
89. C. M. Miranda. *Partial Differential Equations of Elliptic Type.* Springer, 2nd edition, 1970.
90. S. A. Nazarov and B. A. Plamenevsky. *Elliptic problems in domains with piecewise smooth boundaries.* de Gruyter, Berlin, New York, 1994.
91. D. A. Newton and H. Tottenham. Boundary value problems in thin shallow shells of arbitrary plan form. *J. Engng. Math.* **2**, 211–223, 1968.
92. R. Nordgren. On the method of Green's function in the thermoelastic theory of shallow shells. *Int. J. Engng. Sci.* **1**, 279–308, 1963.
93. N. Ortner. Construction of fundamental solutions. In C. A. Brebbia, editor, *Topics in Boundary Element Research.* Springer, 1984.
94. Xiaolin Peng and Guangqian He. Computation of fundamental solutions of the boundary element method for shallow shells. *Applied Mathematical Modelling* **10**, 185–189, 1986.
95. Lu Pin and Huang Maoguang. Computation of the fundamental solution for shallow shells involving shear deformation. *Int. J. Solids Structures* **28**, 631–645, 1991.
96. Lu Pin and Huang Maoguang. Boundary element analysis of shallow shells involving shear deformation. *Int. J. Solids Structures* **29**, 1273–1282, 1992.
97. A. Pomp. A linear, first order shell model for transverse isotropic material under a uniform change of temperature. *Math. Meth. Appl. Sci.* **19**, 1177–1197, 1996.
98. S. Prössdorf. *Some Classes of Singular Equations.* North-Holland, Amsterdam, 1978.
99. P. Rabinowitz and N. Richter. New error coefficients for estimating quadrature errors for analytic functions. *Math. Comp.* **24**, 561–570, 1970.
100. E. Reissner. Stresses and small displacements of shallow spherical shells– parts I and II. *J. of Math. and Phys.* **25**, 80–85 and 279–300, 1946.
101. E. Reissner. On the determination of stresses and displacements for unsymmetric deformation of shallow spherical shells. *J. of Math. and Physics* **38**, 16–35, 1959.
102. D. Ren and K.-C. Fu. A boundary element technique on the solution of cylindrical shell bending problems. In C. A. Brebbia and G. S. Gipson, editors, *Boundary Elements XIII*, pages 499–510, 1991.
103. P. Ricci. Sui potenziale di simplice strato per le equazioni ellitiche di ordne superiore in due variabli. *Rend. Math.*, **7**, pages 1–39, 1974.
104. J. R. Rice. On the degree of convergence of nonlinear spline approximation. In I. J. Schoenberg, editor, *Approximations with special emphasis on spline functions*, pages 349–366, New York, London, 1969. Academic Press.
105. K. Ruotsalainen. On the boundary element method with mesh refinement on curves with corners. *Journ. of Comput. and Appl. Math.* **20**, 373–378, 1987.
106. J. L. Sanders Jr. Cutouts in shallow shells. *J. Appl. Mech., Transactions of ASME* **37**, 374–383, 1970.
107. J. L. Sanders Jr. Singular solution to the shallow shell equations. *J. Appl. Mech., Transactions of ASME* **37**, 361–366, 1970.

108. J. L. Sanders Jr. and J. G. Simmonds. Concentrated forces on shallow cylindrical shells. *J. Appl. Mech., Transactions of ASME* **37**, 367–373, 1970.

109. S. A. Sauter and A. Krapp. On the effect of numerical integration in the Galerkin boundary element method. *Numer. Math.* **74**, 337–359, 1996.

110. C. Schwab. Hierarchic models of elliptic boundary value problems on thin domains, a-posteriori error estimation and Fourier analysis. Habilitation thesis, University of Stuttgart, 1994.

111. C. Schwab. Boundary layer resolution in hierarchical plate modelling. *Math. Meth. Appl. Sci.* **18**, 345–370, 1995.

112. C. Schwab and W. L. Wendland. On numerical cubatures of singular surface integrals in boundary element methods. *Numer. Math.* **62**, 343–369, 1992.

113. L. Schwarz. *Théorie des distributions.* Hermann, Paris, 1966.

114. R. Seeley. Topics in pseudo-differential operators. In L. Nirenberg, editor, *Pseudo-Differential operators*, Rom, 1969.

115. B. A Shoikhet. On asymptotically exact equations of thin plates of complex structure. *PMM, J. Appl. Math. and Mech.* **37**, 867–877, 1973.

116. B. A Shoikhet. An energy identity in physically nonlinear elasticity and error estimates of the plate equation. *PMM, J. Appl. Math. and Mech.* **40**, 291–301, 1976.

117. J. G. Simmonds. Green's functions for closed elastic spherical shells, exact and accurate asymptotic solutions. *Proc. of the Koninklije Nederlandse Academie van Wetenschappen* **71**, 236–249, 1968.

118. J. G. Simmonds and M. R. Bradley. The fundamental solution for a shallow shell with an arbitrary quadratic midsurface. *J. Appl. Mech., Transactions of ASME* **43**, 286–290, 1976.

119. J. G. Simmonds and C. Tropf. The fundamental (normal point load) solution for a shallow hyperbolic paraboloid shell. *SIAM J. Appl. Math.* **27**, 102–120, 1974.

120. N. Simos and A. M. Sadegh. An indirect boundary integral equation to non-shallow spherical shell problems with arbitrary boundary constraints. *J. Appl. Mech., Transactions of ASME* **56**, 918–925, 1989.

121. J. Sládek, V. Sládek and V. Markechová. An advanced boundary element method for elastic problems in nonhomogeneous media. *Acta Mechanica* **97**, 71–90, 1993.

122. E. Stephan and W. Wendland. Remarks to Galerkin and least square methods with finite elements for general elliptic problems. *Manuscripta Geodedica* **1**, 93–123, 1976.

123. A. Tepavitcharov and G. Gospodinov. The BIEM applied to static analysis of thin elastic plates on elastic foundation and thin shallow spherical shells. In *Proc. of Bulg. Math. Soc.*, 1983.

124. Matsui Tetsuya and Matsuoka Osamu. The fundamental solution in the theory of shallow shells. *Int. J. Solids Structures* **14**, 971–986, 1978.

125. N. Tosaka and S. Miyake. A boundary integral equation formulation for elastic shallow shell bending problems. In C. A. Brebbia, editor, *Proc. of 5th Simp. on BEM*, pages 527–538, Japan, 1983.

126. N. Tosaka and S. Miyake. In Du Qinghua, editor, *Boundary Elements*, pages 59–66. Pergamon Press, 1986.

127. H. Tottenham. Boundary element method for plates and shells. In P. K. Banerjee and R. Butterfield, editors, *Developments in Boundary Element Method-1*, pages 173–205, 1979.

128. N. V. Urbanovitch and G. N. Chernyshev. Local stress in a shell resulting from a concentrated load or heat source (in russian). *Isvestija Akademii Nauk SSR, Mekhanika Tverdogo Tela*, Nr. 2, 83–93, 1970.

129. I. N. Vekua. *Generalized Analytic Functions*. Pergamon Press, Oxford, 1962.
130. P. M. Velitchko and V. P. Shevtchenko. On the action of concentrated forces and moments on shells of positive curvature (in russian). *Isvestija Akademii Nauk SSR, Mekhanika Tverdogo Tela*, Nr. 2, 1969.
131. W. Walter. An elementary proof of the Cauchy-Kowalevsky-theorem. *American Math. Monthly* **92**, 115–126, 1985.
132. Jianguo Wang and K. Schweizerhof. Computation of fundamental solutions for laminated anisotropic shallow shells. *Mechanics Research Comm.* **22**, 393–400, 1995.
133. W. L. Wendland. An integral equation method for generalized analytic functions. In *Constructive and Computational Methods for Differential and Integral Equations*, Lecture Notes in Mathematics 430, pages 414–452. Springer, 1974.
134. W. L. Wendland. *Elliptic Systems in the Plane*. Pitman, London, 1979.
135. W. L. Wendland. Strongly elliptic boundary integral equations. In *The state of the art in numerical analysis*. (eds. A. Iserles and M. Powell), Clarendon Press, Oxford, pages 511–561, 1987.
136. J. P. Wilkinson and A. Kalnins. Deformation of open spherical shells under arbitrary located concentrated loads. *J. Appl. Mech., Transactions of ASME* **87**, 305–312, 1966.
137. Tianqi Ye. Analysis of shallow shells by boundary elements. In Q. H. Du and M. Tonaka, editors, *Theory and Application of BEM*, pages 527–538. Springer, 1990. Proc. of 2nd China-Jap. Symp. on BEM, Beijing, China.
138. J. W. Young. *Singular solutions for non-shallow dome shells*. PhD thesis, Stanford University, 1963.

Printing: Druckhaus Beltz, Hemsbach
Binding: Buchbinderei Schäffer, Grünstadt

Lecture Notes in Mathematics

For information about Vols. 1–1490
please contact your bookseller or Springer-Verlag

Vol. 1581: D. Bakry, R. D. Gill, S. A. Molchanov, Lectures on Probability Theory. Editor: P. Bernard. VIII, 420 pages. 1994.

Vol. 1582: W. Balser, From Divergent Power Series to Analytic Functions. X, 108 pages. 1994.

Vol. 1583: J. Azéma, P. A. Meyer, M. Yor (Eds.), Séminaire de Probabilités XXVIII. VI, 334 pages. 1994.

Vol. 1584: M. Brokate, N. Kenmochi, I. Müller, J. F. Rodriguez, C. Verdi, Phase Transitions and Hysteresis. Montecatini Terme, 1993. Editor: A. Visintin. VII. 291 pages. 1994.

Vol. 1585: G. Frey (Ed.), On Artin's Conjecture for Odd 2-dimensional Representations. VIII, 148 pages. 1994.

Vol. 1586: R. Nillsen, Difference Spaces and Invariant Linear Forms. XII, 186 pages. 1994.

Vol. 1587: N. Xi, Representations of Affine Hecke Algebras. VIII, 137 pages. 1994.

Vol. 1588: C. Scheiderer, Real and Étale Cohomology. XXIV, 273 pages. 1994.

Vol. 1589: J. Bellissard, M. Degli Esposti, G. Forni, S. Graffi, S. Isola, J. N. Mather, Transition to Chaos in Classical and Quantum Mechanics. Montecatini Terme, 1991. Editor: S. Graffi. VII, 192 pages. 1994.

Vol. 1590: P. M. Soardi, Potential Theory on Infinite Networks. VIII, 187 pages. 1994.

Vol. 1591: M. Abate, G. Patrizio, Finsler Metrics – A Global Approach. IX, 180 pages. 1994.

Vol. 1592: K. W. Breitung, Asymptotic Approximations for Probability Integrals. IX, 146 pages. 1994.

Vol. 1593: J. Jorgenson & S. Lang, D. Goldfeld, Explicit Formulas for Regularized Products and Series. VIII, 154 pages. 1994.

Vol. 1594: M. Green, J. Murre, C. Voisin, Algebraic Cycles and Hodge Theory. Torino, 1993. Editors: A. Albano, F. Bardelli. VII, 275 pages. 1994.

Vol. 1595: R.D.M. Accola, Topics in the Theory of Riemann Surfaces. IX, 105 pages. 1994.

Vol. 1596: L. Heindorf, L. B. Shapiro, Nearly Projective Boolean Algebras. X, 202 pages. 1994.

Vol. 1597: B. Herzog, Kodaira-Spencer Maps in Local Algebra. XVII, 176 pages. 1994.

Vol. 1598: J. Berndt, F. Tricerri, L. Vanhecke, Generalized Heisenberg Groups and Damek-Ricci Harmonic Spaces. VIII, 125 pages. 1995.

Vol. 1599: K. Johannson, Topology and Combinatorics of 3-Manifolds. XVIII, 446 pages. 1995.

Vol. 1600: W. Narkiewicz, Polynomial Mappings. VII, 130 pages. 1995.

Vol. 1601: A. Pott, Finite Geometry and Character Theory. VII, 181 pages. 1995.

Vol. 1602: J. Winkelmann, The Classification of Three-dimensional Homogeneous Complex Manifolds. XI, 230 pages. 1995.

Vol. 1603: V. Ene, Real Functions – Current Topics. XIII, 310 pages. 1995.

Vol. 1604: A. Huber, Mixed Motives and their Realization in Derived Categories. XV, 207 pages. 1995.

Vol. 1605: L. B. Wahlbin, Superconvergence in Galerkin Finite Element Methods. XI, 166 pages. 1995.

Vol. 1606: P.-D. Liu, M. Qian, Smooth Ergodic Theory of Random Dynamical Systems. XI, 221 pages. 1995.

Vol. 1607: G. Schwarz, Hodge Decomposition – A Method for Solving Boundary Value Problems. VII, 155 pages. 1995.

Vol. 1608: P. Biane, R. Durrett, Lectures on Probability Theory. Editor: P. Bernard. VII, 210 pages. 1995.

Vol. 1609: L. Arnold, C. Jones, K. Mischaikow, G. Raugel, Dynamical Systems. Montecatini Terme, 1994. Editor: R. Johnson. VIII, 329 pages. 1995.

Vol. 1610: A. S. Üstünel, An Introduction to Analysis on Wiener Space. X, 95 pages. 1995.

Vol. 1611: N. Knarr, Translation Planes. VI, 112 pages. 1995.

Vol. 1612: W. Kühnel, Tight Polyhedral Submanifolds and Tight Triangulations. VII, 122 pages. 1995.

Vol. 1613: J. Azéma, M. Emery, P. A. Meyer, M. Yor (Eds.), Séminaire de Probabilités XXIX. VI, 326 pages. 1995.

Vol. 1614: A. Koshelev, Regularity Problem for Quasilinear Elliptic and Parabolic Systems. XXI, 255 pages. 1995.

Vol. 1615: D. B. Massey, Lê Cycles and Hypersurface Singularities. XI, 131 pages. 1995.

Vol. 1616: I. Moerdijk, Classifying Spaces and Classifying Topoi. VII, 94 pages. 1995.

Vol. 1617: V. Yurinsky, Sums and Gaussian Vectors. XI, 305 pages. 1995.

Vol. 1618: G. Pisier, Similarity Problems and Completely Bounded Maps. VII, 156 pages. 1996.

Vol. 1619: E. Landvogt, A Compactification of the Bruhat-Tits Building. VII, 152 pages. 1996.

Vol. 1620: R. Donagi, B. Dubrovin, E. Frenkel, E. Previato, Integrable Systems and Quantum Groups. Montecatini Terme, 1993. Editors:M. Francaviglia, S. Greco. VIII, 488 pages. 1996.

Vol. 1621: H. Bass, M. V. Otero-Espinar, D. N. Rockmore, C. P. L. Tresser, Cyclic Renormalization and Auto-morphism Groups of Rooted Trees. XXI, 136 pages. 1996.

Vol. 1622: E. D. Farjoun, Cellular Spaces, Null Spaces and Homotopy Localization. XIV, 199 pages. 1996.

Vol. 1623: H.P. Yap, Total Colourings of Graphs. VIII, 131 pages. 1996.

Vol. 1624: V. Brınzanescu, Holomorphic Vector Bundles over Compact Complex Surfaces. X, 170 pages. 1996.

Vol.1625: S. Lang, Topics in Cohomology of Groups. VII, 226 pages. 1996.

Vol. 1626: J. Azéma, M. Emery, M. Yor (Eds.), Séminaire de Probabilités XXX. VIII, 382 pages. 1996.

Vol. 1627: C. Graham, Th. G. Kurtz, S. Méléard, Ph. E. Protter, M. Pulvirenti, D. Talay, Probabilistic Models for Nonlinear Partial Differential Equations. Montecatini Terme, 1995. Editors: D. Talay, L. Tubaro. X, 301 pages. 1996.

Vol. 1628: P.-H. Zieschang, An Algebraic Approach to Association Schemes. XII, 189 pages. 1996.

Vol. 1629: J. D. Moore, Lectures on Seiberg-Witten Invariants. VII, 105 pages. 1996.

Vol. 1630: D. Neuenschwander, Probabilities on the Heisenberg Group: Limit Theorems and Brownian Motion. VIII, 139 pages. 1996.

Vol. 1631: K. Nishioka, Mahler Functions and Transcendence.VIII, 185 pages.1996.